Comprehensive Virology 5

Comprehensive Virology

Edited by Heinz Fraenkel-Conrat
University of California at Berkeley

and Robert R. Wagner
University of Virginia

Volume 1: *Descriptive Catalogue of Viruses* – by Heinz Fraenkel-Conrat

Reproduction

Volume 2: *Small and Intermediate RNA Viruses* – Contributors: J.T. August, L. Eoyang, A. Siegel, V. Hariharasubramanian, L. Levintow, E. R. Pfefferkorn, D. Shapiro, and W. K. Joklik

Volume 3: *DNA Animal Viruses* – Contributors: L. Philipson, U. Lindberg, J.A. Rose, N.P. Salzman, G. Khoury, B. Moss, B. Roizman, and D. Furlong

Volume 4: *Large RNA Viruses* – Contributors: P.W. Choppin, R.W. Compans, R.R. Wagner, and J.P. Bader

In preparation:

DNA Bacteriophages – Contributors: D.T. Denhardt, C.K. Mathews, and R.S. Ray

Structure and Assembly

Volume 5: *Virions, Pseudovirions, and Intraviral Nucleic Acids* – Contributors: T.I. Tikchonenko, John T. Finch, Lionel V. Crawford, and H. Vasken Aposhian

Volume 6: *Assembly of Small RNA Viruses* – Contributors: L. Hirth, P.P. Hung, T. Ohno, Y. Okada, K.E. Richards, R.R. Rueckert, and R.C. Williams

Other contributors include: W. Bauer, D.L.D. Caspar, P.W. Choppin, R.W. Compans, F.A. Eiserling, W. Fiers, R.M. Franklin, H.S. Ginsberg, K.A. Harrap, J. King, T.W. Tinsley, J. Vinograd, and W.B. Wood

Regulation and Genetics

Volume 7: *Bacteriophages* – Contributors: R. Calendar, E. P. Geiduschek, M.E. Gottesman, S. Gottesman, and R. Weisberg

Other contributors include: J. Atabekov, D. Baltimore, D. Bishop, M. A. Bratt, A. Campbell, P. Cooper, W. Doerfler, W. Eckhart, B. Fields, H. Ginsberg, H. Hanafusa, A. Huang, E.M.J. Jaspars, J. Kates, A. Lewis, D. Nathans, E.R. Pfefferkorn, C. Pringle, J.R. Roberts, H.L. Sänger, A.J. Shatkin, F. W. Stahl, J.H. Subak-Sharpe, W. Szybalski, L. Van Vloten-Doting, and P. Vogt

Interaction of Viruses and Their Hosts

Effects of Physical and Chemical Agents

Comprehensive

Edited by
Heinz Fraenkel-Conrat
Department of Molecular Biology and Virus Laboratory
University of California, Berkeley, California

and
Robert R. Wagner
Department of Microbiology
University of Virginia, Charlottesville, Virginia

Virology

5

Structure and Assembly

Virions, Pseudovirions, and Intraviral Nucleic Acids

PLENUM PRESS • NEW YORK AND LONDON

Library of Congress Cataloging in Publication Data

Fraenkel-Conrat, Heinz, 1910-
Comprehensive virology.

Includes bibliographies.
1. Virology—Collected works. I. Wagner, Robert R., 1923- joint author.
II. Title. [DNLM: 1. Virus diseases. 2. Viruses. QW160 F799ca]
QR357.F72 576'.64'08 74-5494
ISBN 978-1-4684-2711-0 ISBN 978-1-4684-2709-7 (eBook)
DOI 10.1007/978-1-4684-2709-7

Softcover reprint of the hardcover 1st edition 1975
A Division of Plenum Publishing Corporation
227 West 17th Street, New York, N.Y. 10011

United Kingdom edition published by Plenum Press, London
A Division of Plenum Publishing Company, Ltd.
Davis House (4th Floor), 8 Scrubs Lane, Harlesden, London, NW10 6SE, England

Foreword

The time seems ripe for a critical compendium of that segment of the biological universe we call viruses. Virology, as a science, having passed only recently through its descriptive phase of naming and numbering, has probably reached that stage at which relatively few new—truly new—viruses will be discovered. Triggered by the intellectual probes and techniques of molecular biology, genetics, biochemical cytology, and high-resolution microscopy and spectroscopy, the field has experienced a genuine information explosion.

Few serious attempts have been made to chronicle these events. This comprehensive series, which will comprise some 6000 pages in a total of about 22 volumes, represents a commitment by a large group of active investigators to analyze, digest, and expostulate on the great mass of data relating to viruses, much of which is now amorphous and disjointed, and scattered throughout a wide literature. In this way, we hope to place the entire field in perspective, and to develop an invaluable reference and sourcebook for researchers and students at all levels.

This series is designed as a continuum that can be entered anywhere, but which also provides a logical progression of developing facts and integrated concepts.

Volume 1 contains an alphabetical catalogue of almost all viruses of vertebrates, insects, plants, and protists, describing them in general terms. Volumes 2–4 deal primarily, but not exclusively, with the processes of infection and reproduction of the major groups of viruses in their hosts. Volume 2 deals with the simple RNA viruses of bacteria, plants, and animals; the togaviruses (formerly called arboviruses), which share with these only the feature that the virion's RNA is able to act as messenger RNA in the host cell; and the reoviruses of animals and plants, which all share several structurally singular features, the most important being the double-strandedness of their multiple RNA molecules. This grouping, of course, has only slightly more in its favor than others that could have been, or indeed were, considered.

Volume 3 addresses itself to the reproduction of all DNA-containing viruses of vertebrates, a seemingly simple act of classification, even though the field encompasses the smallest and the largest viruses known. (The reproduction of the DNA-bacteriophages will be dealt with later, probably in Volume 8.)

The reproduction of the larger and more complex RNA viruses is the subject matter of Volume 4. These viruses share the property of lipid-rich envelopes with the togaviruses included in Volume 2. They share as a group, and with the reoviruses, the presence of enzymes in their virions and the need for their RNA to become transcribed before it can serve messenger functions.

Volumes 5 and 6 represent the first in a series that focuses primarily on the structure and assembly of virus particles. Other volumes nearing completion will deal with animal virus genetics, multicomponent plant viruses, regulation of viral replication, virus–host relationships, etc. The completed series will endeavor to encompass all aspects of the molecular biology and the behavior of viruses.

We hope to keep this series up to date at all times by prompt and rapid publication of all contributions, and by encouraging the authors to update their chapters by additions or corrections whenever a volume is reprinted.

Contents

Chapter 1

Structure of Viral Nucleic Acids *in Situ*

T. I. Tikchonenko

Chapter 2

Structure of Small DNA-Containing Animal Viruses

John T. Finch and Lionel V. Crawford

Chapter 3

Pseudovirions in Animals, Plants, and Bacteria

H. Vasken Aposhian

CHAPTER 1

Structure of Viral Nucleic Acids *in Situ*

T. I. Tikchonenko

Institute of Virology
USSR Academy of Medical Sciences and
Department of Virology
Moscow State University
Moscow, USSR

The main principles of the structural organization of nucleic acids in solution having been established, scientists are increasingly turning their attention to the fine features of the secondary structure of DNA and RNA and the possibilities of their reversible conformational alterations under the influence of various external factors. These studies have contributed toward an understanding of the fact that the environments of nucleic acids in ribo- and deoxyribonucleoproteins (RNP and DNP) of various origin differ significantly from their environments in the experimental test tube. Thus, studies of conformation of the nucleic acids *in situ* appear to be the order of the day.

With regard to viruses, it is reasonable to consider this problem separately for isometric and filamentous, or rodlike, viruses, as the conditions of interaction of RNA and DNA with protein in these two large groups of viruses differ greatly. In the isometric virion only a limited amount of nucleic acid interacts with protein and, as a rule, it involves only the peripheral parts of nucleic acid, whereas in the filamentous viruses the whole length of the nucleic acid molecule is literally buried in protein.

The peculiarity of the structure of viral particles, their rigid geometric organization, and their characteristics of biogenesis lead to the fact that the structure of their nucleic acids *in situ* in closely related to the structure of the protein capsid. This is particularly evident in the filamentous and rodlike viruses of plants, the proteins of which are capable of self-assembly (Schramm and Zillig, 1955).

1. VIRUSES WITH SINGLE-STRANDED NUCLEIC ACIDS

1.1. Rodlike and Filamentous RNA-Containing Viruses

The macromolecular structure of single-stranded RNA has been most extensively studied in particles of the tobacco mosaic virus (TMV), the potato X virus (PXV), and other similar viruses from the genera *Tobamovirus* and *Potexvirus*.

1.1.1. RNA Packing *in Situ*

As evidenced by the results of X-ray diffraction analysis and electron microscopy, the TMV particle is a cylinder with an external diameter of 15–18 nm, a length of 300 nm, and an internal channel 4 nm in diameter (Franklin, 1955*a,b,* 1956*a,b*; Franklin and Klug, 1956; Caspar, 1956, 1963; Casper and Klug, 1963; Hart, 1955; Huxley and Zubay, 1961; Klug and Caspar, 1960). The protein capsid is formed by 2130 elipsoid-shaped polypeptide subunits of 17,500 molecular weight, containing 158 amino acids (Fraenkel-Conrat, 1969). The capsid has a helical-type symmetry—the pitch of the helix is 2.3 nm, the period of identity (three turns) is 6.9 nm, and the total number of turns is 130. Thus, there are $16\frac{1}{3}$ protein subunits per helix turn and 49 subunits per period of identity.

The X-ray diffraction studies of the whole TMV particles and their protein aggregates without RNA have demonstrated that the RNA strand is deeply buried in protein, following the pitch of the protein helix. A comparison of the radial distribution of the electron density in TMV and in the protein aggregates clearly indicates that the RNA is located in the virion at a distance of 4 nm from the axis of the particle. Thus, the tertiary structure of TMV RNA *in situ* may be described as a rigid, rodlike helix with a diameter of 8 nm and a pitch of 2.3 nm. The minimum distance of the RNA strand from the external surface of the particle is 3.5 nm, and it is 2.0 nm from the internal surface of the central channel. As the molecular weight of TMV RNA is 2.05×10^6 (Boedtker, 1968), there are 49 nucleotides per turn of the RNA helix, or 3 nucleotides per protein subunit. As calculated by Ginoza (1958), the single-stranded polynucleotide chain with the molecular weight of 2×10^6 and the total length of 3300 nm may form a helix possessing the above parameters provided that the distance between the neighboring nucleotides in the helix is 0.5 nm.

The same structure was found for the closely related cucumber mosaic virus (CMV) (Franklin *et al.*, 1957; Ginoza, 1958; Holmes and Franklin, 1958; Kilkson, 1957) and for other rodlike viruses of plants with helical symmetry, e.g., tobacco rattle virus and barley stripe virus (Offord, 1966; Finch, 1965). The number and size of the protein subunits in CMV are also identical to those of TMV (Hill and Shepherd, 1972; Tung and Knight, 1972). Slight differences in the electron density distribution observed between the X-ray diffraction patterns of CMV and TMV in the region directly adjoining the internal channel of the particles are yet to be interpreted (Caspar and Klug, 1963), even though these very differences in the capsid structure may determine the difference in the reactivities between the RNA bases in TMV and CMV (see below). Nevertheless, owing to the similarity of composition and structure of the protein capsid, the RNA conformation *in situ* in these two viruses should be quite similar.

Filamentous PXV is also a hollow cylinder 580 nm in length and 13 nm in diameter with an internal channel 3.4 nm in diameter (Bode and Paul, 1955; Reichmann, 1959; Tollin *et al.*, 1967; Varma *et al.*, 1968; Wilson and Tollin, 1969). The above dimensions relate to the viral preparations under 100% humidity since the particle is observed to contract upon drying, giving rise to differences in values reported by various authors. The PXV capsid is formed of 1475 elliptically shaped protein subunits which have a molecular weight of 22,300. The pitch of the protein helix, as reported by various authors, is 3.3–3.6 nm, corresponding to 160–175 turns. There are 9–10 protein subunits per turn of the helix (Brandes and Bercks, 1965; Miki and Knight, 1968; Tollin *et al.*, 1967).

Similar to TMV, the RNA of PXV is deeply buried in the protein capsid at a distance of 3.5 nm from the center of the particle, forming the helix corresponding to packing of protein subunits (Varma *et al.*, 1968; Wilson and Tollin, 1969). The distances between the RNA strand and the external surface of the capsid and the surface of the internal channel are 3 and 2.1 nm, respectively. Such a helix is about 6 nm in diameter and the distance between the neighboring nucleotides in the helix is 0.55 nm (Wilson and Tollin, 1969). As the molecular weight of RNA of PXV is 2.1×10^6 and there are 160–175 turns, one turn of the helix contains 38–42 nucleotides. In this case there are 4 nucleotides per protein subunit (Wilson and Tollin, 1969).

A similar, though not identical, structure has been found in viruses belonging to the Y subgroup of the potato viruses; among these, the turnip mosaic virus has been most extensively studied (Varma *et al.*, 1968). The length of the nucleocapsid in this virus is 790 nm, its diameter is 15 nm, and the pitch of the protein helix is 3.4 nm.

Given in Table 1 are the helix parameters for the rodlike TMV and the filamentous PXV. As may be seen, the differences between the helix parameters for intraviral RNAs of these two viruses are rather significant, exceeding by far those alterations that accompany conformational changes of the secondary structure of various RNAs and synthetic polyribonucleotides (Gomatos *et al.*, 1964; Tomita and Rich, 1964; Langridge and Gomatos, 1963; Harrison *et al.*, 1971).

None of the natural RNAs or polyribonucleotides has been observed to give in solution a helix with such geometry. By comparison, the helical part of the single-stranded RNAs has the A-configuration geometry (Zubay and Wilkins, 1960; Fuller, 1961; Brown and Zubay, 1960). In this case the pitch of the helix is 2.9 nm, the diameter is 1.7 nm with 11 nucleotides per helix turn, and the distance between nucleotides is 0.25 nm (Tomita and Rich, 1964; Gomatos *et al.*, 1964). This inconsistency is readily explained by the fact that the RNAs of these viruses *in situ* have no inherent secondary structure and assume the secondary structure of the protein helix owing to close interaction with protein. Indeed, if we treat the secondary structure of a polymer as though the helical organization and the periodicity are based on the "close-range interaction," i.e., on the interaction between the neighboring monomers and chain links (Spirin, 1963), then the intraviral RNAs

TABLE 1

Geometry of the RNA Helix in Cylindrical Plant Viruses[a]

Virus	Diameter	Pitch	Length	Number of turns	Number of nucleotides per turn	Distance between nucleotides[b]
Tobacco mosaic virus	8	2.3	300	130	49	0.5
Potato X virus	6	3.3–3.6	580	160–175	38–42	0.55

[a] All values in nm.
[b] The distance between the neighboring nucleotides along the helix.

have no inherent secondary structure and cannot have one. As the spatial organization of RNA in TMV and PXV is distinguished by a high degree of order and stability, and the absence of the regular secondary structure is conventionally associated with disorder and chaos, many authors prefer to speak in these cases about the "formal secondary structure" to eliminate misunderstanding. The parameters of the RNA structure *in situ* given in Table 1 indicate quite clearly that with such a helix diameter and with such distances between the nucleotides, H bonding of the bases is not possible either within one turn (perpendicular to the long axis of the particle) or between different turns (parallel to the long axis of the particle). The intrastacking interactions are not understood as clearly. The conventional viewpoint put forward first by Franklin *et al.* (1959) and extended later by Caspar and Klug (1963) is that the intrastacking interactions of the bases in intraviral TMV RNA are absent owing to the maximum distances between nucleotides and anomalous arrangement of the planes of the bases with respect to the long axis of the RNA molecule. Subsequently this viewpoint came to be generally accepted, and it is only recently that some doubts have appeared regarding its validity (Cheng, 1968; Bush and Sheraga, 1967).

1.1.2. Interaction between the Bases in RNA *in Situ*

The absence of complementary and, most probably, intrastacking interaction of the bases in RNA in the rodlike virions must be due to the peculiarities of RNA packing *in situ* discussed above. At present, there are a number of independent results confirming this viewpoint, though not always unambiguously. The most convincing confirmation has been furnished by UV spectrophotometry and chemical modification, primarily, by means of nitrosoguanidine. The initial unlimited optimism concerning the capability of birefringence and UV flow dichroism methods to provide accurate quantitative data on the arrangement of bases in RNA *in situ* has given way to some extent to a greater caution. All the authors using optical rotatory dispersion (ORD) and circular dichroism (CD) methods to study the structure of intraviral RNA agree that the spectral evidence obtained indicates the absence of complementary interactions in TMV RNA. There is a divergence of views concerning the intrastacking interactions—some workers believing that interactions of this type exist, while the others adhere to the contrary viewpoint (see below).

As the biophysicists have been interested in TMV for a comparatively long time, the optical studies on TMV already have a relatively long history, going back to the 1950s, including repeated revisions and verifications stemming from theoretical developments, improvement of the experimental techniques, and the general progress of our knowledge in this field.

Evidently, in the absence of intrastacking and complementary interactions in the RNA, hypochromism should be absent, i.e., the intraviral RNA should possess maximum hyperchromism and the same high specific absorption as the denatured preparations. In 1954, Fraenkel-Conrat (1954), studying the interaction between TMV and formaldehyde, demonstrated that incubation with 1–2% HCHO does not change the E_{260} value, that is, the intraviral RNA is not hypochromic. Later these experiments were repeated to take into account the contributions of light scattering and the inherent absorption of protein (Dobrov *et al.*, 1972*b*). It became apparent that although incubation with HCHO did not change the E_{260} value in a suspension of TMV, the value of E_{280} was not altered either, even though the hydroxymethylation of bases should be accompanied by an increase in absorption in the long-wavelength part of the spectrum at 270–290 nm (Grossman *et al.*, 1961; Haselkorn and Doty, 1961). At the same time, incubation of the closely related CMV and PXV with HCHO is accompanied by the expected increase in absorption at 280 nm without the respective increase in the value of E_{260} (Dobrov *et al.*, 1972*b*). Calculations have shown that the specific absorption of RNA in TMV, CMV, and PXV is the same and is characteristic of the complete hyperchromism which is typical of the denatured state of RNA.

The amount of hyperchromism at 260 nm for the intraviral RNA is 23% for TMV and PXV and 28% for CMV (Dobrov *et al.*, 1973, 1974), and a value of 25.5% has been found for the turnip mosaic and tobacco etch viruses belonging to the Y group of potato viruses (Damirdagh and Shepherd, 1970; Hill and Shepherd, 1972). These values correspond, on average, to the limiting hyperchromism for RNA preparations denaturated *in vitro* (Spirin, 1963). The value of the hyperchromic state of RNA in CMV somewhat exceeds this average value, but the reasons for that are as yet unknown. In principle, increased hyperchromism is either associated with a disturbance of the intrastacking interactions or is due to unknown individual properties of CMV preparations. Unfortunately, studies of this problem are greatly hindered by the procedural difficulties in the determination of the specific RNA absorption in the virions of the rodlike viruses with low nucleic acid content.

At the same time, the amino groups of the bases in TMV are quite inaccessible to HCHO owing to their interaction with protein. When

TMV, CMV, and PXV are disrupted, the RNA undergoes conformational transition and acquires the usual secondary structure (Bonhoeffer and Schachman, 1960; Dobrov *et al.*, 1972). As expected, this process is accompanied by a pronounced hypochromic effect at favorable values of *p*H and ionic strength. However, the first determinations of this type (Bonhoeffer and Schachman, 1960) did not take into account the protein contribution to the total absorption of the preparation and the significant alteration of its value after denaturation leading to an underestimated value of the RNA hypochromism. Thus, the completely hyperchromic state of the RNA in the particles of the three viruses studied indicate the absence of fixed base stacking *in situ*.

This view is confirmed, to a certain extent, by the results of the ORD studies of TMV (Bush and Sheraga, 1967; Cheng, 1968; Simmons and Glazer, 1966). The ORD spectra of the intraviral RNA in this case are determined from the difference between the spectra of the intact virus and of the repolymerized viral protein, implicitly assuming that the contribution of protein in the ORD of both intact virus and empty protein capsid is the same. There is another method for determination of the ORD spectra of the intraviral RNA which is based on the difference between the rotations of the intact and disrupted viruses and can be used provided that the RNA is not renaturated after virus disruption (Kust *et al.*, 1972). This approach proceeds from the absence of significant differences between the rotations of the A protein and of the repolymerized protein in the spectral region above 260 nm (Cheng, 1968; Simmons and Blaut, 1960), and its validity is confirmed by the similarity of the differential spectra in the 270–300 nm range for the intraviral RNA obtained using both procedures (Dobrov *et al.*, 1973).

A combination of these procedures allows the determination of the ORD curves for a series of rodlike and filamentous viruses, including those from which repolymerized protein cannot be obtained (Dobrov *et al.*, 1972*a*, 1973). It has been found that the ORD curves for these viruses differ radically, irrespective of the evaluation procedure for the differential spectra. The ORD spectra for RNAs of TMV and PXV *in situ* are similar to the ORD pattern for the denaturated RNA preparations in distilled water or in the presence of urea. In both cases there is observed a strong positive maximum in the 280–300 nm region, the position and the transition point of which are shifted by 6–8 nm to the long-wavelength side of the spectrum as compared to the RNA spectra in salt solutions. However, along with similarities between spectra of RNA in TMV particles and spectra of the RNA preparations in distilled water, they reveal certain differences. In the intraviral RNA the maximum of positive rotation is shifted further to the long-wavelength side and the amplitude of the positive peak is twice that for this maximum in the spiralized preparations of RNA. So long as only the ORD was known for TMV, interpretation of these data was based on the traditional concepts about certain correlations between the secondary structure of nucleic acids and their optical activity (Cantor *et al.*, 1966; Cantor and Tinoco, 1965; Warshaw and Tinoco, 1965). In accordance with this view it was assumed that the amplitude of the positive peak in the 270–280 nm region has to be attributed mainly to base stacking, while the positions of the rotation maximum and transition point, the other conditions being equal, depend on the complementary interactions of the bases and H bonds between them. According to this, the peculiarity of the ORD spectra for intraviral TMV RNA has been ascribed to the absence of H bonds in RNA under these conditions (Simmons and Glazer, 1966; Taniguchi *et al.*, 1971; Bush and Sheraga, 1967; Cheng, 1968). While the interpretation up to this point corresponds to the available knowledge about the peculiarities of TMV RNA structure *in situ,* it is more difficult to interpret the profound positive maximum which, according to Cheng and Bush and Sheraga should be a result of strong base stacking under the conditions of rigid fixation of the bases (so-called low-temperature form of base stacking).

In this case it should be suggested that base stacking in RNA *in situ* should be considerably more ordered than in TMV RNA in solution, even under conditions of maximum spiralization (over 80% content of formal secondary structure). This conclusion is evidently in conflict with a whole set of data on RNA packing in the particles of the rodlike and filamentous viruses, and it does not agree with the ORD spectra obtained for other viruses of this group, including mutants of TMV. In particular, in CMV, the capsid of which [according to the results of X-ray diffraction studies and electron microscopy (Caspar, 1963)] does not differ from that of TMV, the intraviral RNA does not have positive rotation at all (Dobrov *et al.*, 1972*a*). The same result has been obtained for a strain of TMV, the Dolichos mosaic virus (DMV) (Dobrov *et al.*, 1973). If we argue within the framework of traditional concepts and in terms of formal logic, the radical differences in the ORD spectra of the viruses studied alone should lead us to the no less radical conclusion that in DMV virions the RNA has ordered base stacking and in other TMV strains and CMV there are no intrastacking interactions in the RNA *in situ.* However, such scientific radicalism is noticeably inconsistent with the data about the similar structural organization of viruses in this group in general, as well as with the general concept of the leading role played by the protein capsid in determining the structure of the RNA within the virion. Therefore, it appears preferable to propose that the differences in the ORD spectra of different TMV strains are determined not by such fundamental features of the structural organization of virions as the presence or absence of regular base stacking but by less significant structural peculiarities within the limits of small-scale interstrain differences. Furthermore, we should remember that if the former viewpoint were true we could expect profound biological differences in the properties of TMV, CMV, and their strains, which are not observed in practice. In this case we support Cheng's (1968) idea that the peak of positive rotation of TMV and PXV is due not to the intrastacking interaction of the bases but to a certain orientation of the aromatic amino acids or, more exactly, to interactions ("mixed stack") of the bases with the rings of aromatic amino acids (see below). It cannot be ruled out here that the same effect may be produced by orientation of other optically active components of the polypeptide chain in close partnership with RNA. This view allows for the "large-scale optical effects" accompanying such a common occurrence in the "life" of a virus as the mutation substitution of amino acids in its protein coat. Taking into account that the protein subunits are identical and that mutation gives rise to simultaneous alteration of all 2130 subunits (in a particle of TMV type), we find that replacement of one tryptophan residue (Trp^{52}) by leucine, for example, will result in the disappearance in the virion of the factor responsible for positive rotation.

The occurrence of strong intrastacking interactions in the RNA of rodlike virions *in situ* also does not agree with the inability of UV light to induce formation of pyrimidine dimers in TMV. As is known, the only identified photoproducts in UV-irradiated TMV are the hydrates of uracil and cytosine (Carpenter and Kleczkowski, 1969; Tao *et al.*, 1969). Some authors believe that cross-links are formed between RNA and protein in irradiated TMV (Tao *et al.*, 1966; Goddard *et al.*, 1966), although covalent cross-links have not been directly identified in these experiments.

We shall discuss this problem in more detail in connection with the interaction between RNA and protein, while here it is appropriate to recall the known relationship between formation of dimers, on the one hand, and the secondary structure of the polymer, on the other (Tramer *et al.*, 1969; Hosszu and Rahn, 1967; Kochetkov *et al.*, 1970). Dimerization of pyrimidines is hindered in double-helix structures, increasing after thermal denaturation and sharply falling when solvents disturb the interplane interactions.

The existence of ordered base stacking in the RNA of TMV *in situ* is questioned also by the results of flow birefringence and UV dichroism (Franklin *et al.*, 1959; Bendet and Mayfield, 1967; Caspar and Klug, 1963; Gabler and Bendet, 1972; Schachter *et al.*, 1966, Mayfield and Bendet, 1970*a,b*). Ordered base stacking in polynucleotides results if the planes of the bases are preferentially oriented perpendicular to the long axis of the chain. Under these conditions the planes of the bases prove to be arranged in the form of a stack with an angular displacement (the rotation angle) determined by the conformation of the structure as a whole. In studies of the dichroism of oriented DNA and RNA preparations, a number of authors have demonstrated that the plane of the bases in the nucleic acid helix is oriented perpendicular to the chain axis (Spirin, 1963; Frisman *et al.*, 1963).

In contrast to this, a number of authors believe that the base planes in the TMV particle are oriented "more or less parallel" to the long axis of the particle and RNA molecule (Bendet and Mayfield, 1967; Schachter *et al.*, 1966; Mayfield and Bendet, 1970). This conclusion is based on the positive dichroism over the entire range of UV absorption, with a maximum at 254 nm. Even though this indefinite description "more or less parallel" gave way in some later reports to the exact angle of inclination (θ) of the base plane to the particle axis, the validity of such precise calculations is somewhat doubtful (Gabler and Bendet, 1972). However, before considering them we have to stress that these doubts do not involve the concept of the essentially different orientation of the RNA bases in the TMV particle as compared to their orientation in the free RNA.

Some circumstances which have been discussed in detail earlier (Tikchonenko, 1969) may influence to some extent precise calculations of the angle θ. Four such factors are essential for application of the method of flow UV dichroism to this problem. The first factor is the exact knowledge of the fraction of oriented particles, F, a value which enters into Frazer's equation used for calculating the values of θ. Unfortunately, the value of F was determined in a more or less arbitrary fashion in the earlier studies of TMV, phage fd, and the preparations of nucleic acids mentioned above.

This factor, however, seems to lose its importance after experiments with dichroism measurements in pulsed electric fields (Allen and van Holde, 1971). Asymmetric particles of the TMV type are easily oriented in the applied electric fields so as to maximize the induced moment, i.e., parallel to the applied field strength. Thus, the perfect orientation of TMV particles is achieved in these experiments when saturation of the signal strength is obtained at increasingly high signal strengths. There is every reason to believe that the orientation function under these conditions is equal to 1. In addition to that, Mayfield and Bendet (1970*a*) presented a general method for calculating from experimental data the dichroism that would be observed from a suspension of rodlike particles all rods of which are aligned strictly parallel to one another.

Second, the available technique of flow UV dichroism, even when using monochromatic light, allows one to determine the orientation of all chromophores absorbing UV light rather than just the orientation of the bases alone. In other words, for viruses this method determines the average orientation of the four bases in the RNA molecule and of the three aromatic amino acids in the capsid, mainly tryptophan, because its contribution to the UV protein absorption is the dominant one. Third, in order to allow for the contributions of RNA and protein to the results of measurements in such experiments intact TMV particles and the viruslike particles (empty protein capsids, i.e., repolymerized protein not containing RNA) are used (Schachter *et al.*, 1966; Bendet and Mayfield, 1967). Some workers have used for this purpose the artificial nucleocapsid reconstituted from a mixture of virus protein and synthetic polynucleotides, in this case poly(A) (Gabler and Bendet, 1972). In all these cases it has to be assumed that the orientation of the aromatic amino acids in the empty protein capsids or of the adenine residues in the virions reconstituted from a mixture of protein and poly(A) does not differ from the orientation of the chromophores in the native TMV. As for the second and third factors, there is serious indirect evidence confirming the validity of the assumptions used [see Tikchonenko (1969)].

Finally, the fourth factor is the so-called "form dichroism." Dielectric theory according to Mayfield and Bendet, (1970*b*) predicts that nonspherical absorbing particles will be dichroic even if they are composed of isotropic materials. However, these authors have shown that the UV dichroic spectra of TMV and fd particles are practically the same in solvents differing considerably in their refractive index. Thus, for these viruses at least, form dichroism is much less important than it was postulated earlier.

The above considerations indicate the necessity of some caution when evaluating the angle of inclination of the bases reported by various authors (Bendet and Mayfield, 1967; Gabler and Bendet, 1972; Mayfield and Bendet, 1970*a,b;* Taniguchi *et al.*, 1971). Nevertheless, we may note good agreement between the values of θ calculated for different experimental conditions: 42° is the average value for all chromophores in TMV (Bendet and Mayfield, 1967; Mayfield, 1968);

50.8–51.4° is the value for the bases in TMV reported by Taniguchi *et al.* (1971), and 52° is the value reported for adenine residues in viruslike particles reconstituted from TMV protein and poly(A) (Gabler and Bendet, 1972).

Gabler and Bendet (1972) have reported the spectra of flow UV dichroism for three TMV strains, including the bean strain, which contains only one tryptophan residue in contrast to three residues for wild-type and tomato strains. This work allows one to calculate for the bean strain the angle of inclination of the indole ring with respect to the long axis of the particle. Similar to the bases, this angle is 51°, indicating the same orientation in TMV particles of the bases and aromatic amino acids, at least tryptophan.

The same value for the orientation angle of chromophores in the 270–290 nm region of the spectrum has been obtained in experiments with dichroism in pulsed electric fields (Allen and van Holde, 1971) when orientation of rodlike virions is close to 100%. This means that the first factor (see above) does not in reality influence the calculated value of the orientation angle in the experiments with flow dichroism. At the same time the modest magnitude of the dichroism indicates that while there is some preferential orientation of aromatic amino acids and bases whose ring planes are parallel to the helix axis, this orientation is by no means complete. The value for the orientation angle of chromophores in TMV (51°) is not very far from the value for random orientation (54°45′) (Allen and van Holde, 1971).

The absence of intrastacking interactions in the RNA of TMV *in situ* has been rather convincingly demonstrated by the chemical modification of this virus by means of an alkylating agent, *N*-methyl-*N*′-nitro-*N*-nitrosoguanidine (MNNG). As shown by Singer and Fraenkel-Conrat (1969*a,b*), MNNG reacts with cytidine, guanosine, and adenosine (see Table 2). The first and second reactions yield products identified as 3-methylcytidine (3-MC) and 7-methylguanosine (7-MG) while the third reaction gives rise to three derivatives: 1-, 3-, and 7-methyladenosine (1-MA, 3-MA and 7-MA). Alkylation is characterized by the pronounced dependence on the secondary structure of the polymer and, primarily, on the intrastacking interactions involving the bases being modified. This dependence is most clearly seen for cytidine, where transition from the reaction with monomer to the reaction with single-stranded polymer or even double-stranded polymer leads to strong inhibition of the reaction. In contrast, alkylation of the purines by MNNG is sharply increased for polynucleotides. As the 1 position of the purines is involved in complementary interactions, formation of

TABLE 2

Reactions of Various Bases with Nitrosoguanidine as a Function of Interplane and Complementary Interactions[a]

Modification products	Nucleotides	Polymer: Single-stranded poly(A)	Polymer: Double-stranded poly(AU)	TMV RNA: In water	TMV RNA: In DMF[b]	RNA in TMV
1-MA		92	1.2	9.2	12.2	6.3
3-MA	} 1.5	38 { 2	45	1.3 } 16.8	1.8 } 19.1	4.9 } 17.2
7-MA		7	65	6.3	5.1	6
7-MG	5	15 —	—	76	59	48
3-MC	20	1.5 —	—	7.1	22	35

[a] All the values in the table, with the exception of the first two columns, are the percentages of the given type of modified product with respect to the total amount of the alkylated bases. The values in the first two columns are the percentages of alkylation of the given base. The figures in the braces are the percentages of given adenosine derivatives with respect to the total amount of alkylation (Singer and Fraenkel-Conrat, 1969*a*, *b*).

[b] DMF, dimethylformamide.

1-MA is greatly depressed by H bonding. These results are in good agreement with the results of alkylation of free TMV RNA in water and in dimethylformamide (DMF). It is known that in the first case the complementary interactions are disturbed in RNA but the intrastacking interactions are preserved, while in the second case there occurs complete disappearance of the secondary structure. As should be expected, in DMF as compared to water alkylation of cytidine increases and alkylation of guanosine decreases. There is only a slight difference between alkylation of adenosine in water and in DMF.

The results of the *experimentum crucis* are presented in the last column of Table 2. It has been found that among the products of alkylation of the bases in the intraviral RNA as compared to the free RNA, the variation of the proportions of the individual bases is the same as in the *in vitro* reaction in the presence of DMF. Guanosine and cytidine in the intraviral RNA are alkylated to a similar extent; this is due either to a significant acceleration of modification of cytidine or to a lowered rate of modification of guanosine, or to both effects occurring concurrently. Alkylation of adenosine in the RNA of TMV *in situ* and *in vitro* proceeds more or less similarly.

Analyzing these data, one should note that the absolute rates of the reactions between the bases and MNNG in TMV are extremely low. Unfortunately, the exact values of the reaction rate constants for alkylation of various RNA forms have not been reported, but only the fact that in five days of incubation not more than two bases per virion are modified in TMV. This seems to be due to the limited diffusion through the protein capsid owing to the large size of the reacting molecules. Starowsky (1971) has estimated that for compounds of the MNNG type the size becomes a critical factor in determining the possibility of entering virions. This may lead to MNNG being able to react only with the bases in the first and the last turns of the RNA helix at the ends of the rodlike particle. This assumption may certainly be easily checked in the experiments with labeled MNNG by analyzing the distribution of the label in RNA isolated from the modified virions.

The modification of intact TMV by means of MNNG is also accompanied by peculiar biological effects, in particular, by a sharp shift of the ratio between inactivating and mutagenic actions of this compound. Data given in Table 3 indicate that the ratio between the number of mutants and the number of surviving particles after treatment with MNNG is an order of magnitude greater for the intact virus as compared to the infectious RNA. Singer and Fraenkel-Conrat (1969*a*,*b*, 1970) relate these changes to the relatively increased modification of cytidine residues and the appearance of 3-MC in the intraviral RNA.

A higher level of mutagenesis has been found also for the deamination of the intact TMV as compared to that of the free RNA.

Thus, in the intraviral RNA there are no intrastacking and, most probably, complementary interactions of the bases which are responsible in solution for creation of the ordered secondary structure of polynucleotide. It would seem natural to conclude that the viral

TABLE 3

Mutagenic Action of Various Reagents on TMV RNA in Free State and in Virus[a]

Agent	Free RNA	RNA in TMV
Nitrous acid	34	110
Various alkylnitronitrosoguanidines	2–7	51–68

[a] The mutagenic action is expressed as the ratio between the number of local lesions on *Nicotinia sylvestris* for the modified and unmodified RNA or TMV, the number of infective particles being the same.

RNA *in situ* should be totally disordered, but such a conclusion is quite inconsistent with the above experimental results. In the particles of TMV, PXV, and other similar viruses, the secondary structure is highly ordered and this formal secondary structure comprises practically 100% of the particle. This order is due to interaction between the RNA and protein. The great importance of the interaction between the protein and nucleotides for creation of the ordered structure of the latter has been demonstrated by the experiments of Wagner and Arav (1968). These authors have shown that adsorption of mononucleotides (!) by poly-L-lysine leads to the formation of base stacking and to the appearance of the hypochromism typical of the interaction of the bases in oligonucleotides. Under these conditions, mononucleotides interacting with protein behave as if they are linked with phosphodiester bonds.

1.1.3. Interaction between RNA and Protein

The totality of the data discussed above and primarily the results of studies of flow birefringence and UV dichroism permit certain conclusions to be drawn concerning the peculiarities of RNA conformation in virions and the causes of these peculiarities. If the planes of the bases are "predominantly parallel" to the long axis of the particle, the sugar–phosphate skeleton is perpendicular to the long axis. This location of the RNA helix is in agreement with the geometry of the α helix of the viral protein in the intact virus and in the repolymerized capsids. It is assumed that about 35% of the TMV polypeptide chain has the α-helix structure and the spiralized segments are oriented at a certain angle but are "predominantly perpendicular" with respect to the long axis of the particle (Simmons and Blout, 1960; Fraser, 1952; Caspar, 1956). The side chains of amino acids, aromatic in this case, are, similar to the bases, parallel to the long axis of the particle. A similar arrangement of the base planes and of the aromatic amino acid planes seems to be a condition for their interaction. It should be stressed that the interaction between the bases and protein amino acids in the virions of rodlike viruses, although not yet proved by direct and unequivocal experimental evidence, is absolutely necessary on theoretical grounds, and without it a great number of facts cannot be explained.

In particular, the ORD spectra discussed above are among the data demonstrating the probability of interaction between the amino acids and the bases of intraviral RNA. In contrast to the ORD spectra for RNA in solution, the spectra for RNA in the particles of TMV and PXV have not a broad single peak of positive rotation with the maximum at 288 nm, but a double peak with two maxima, at 293 and 286 nm (Cheng, 1968; Dobrov *et al.*, 1972*a*). The protein of TMV and PXV contains three tryptophan residues (Fraenkel-Conrat, 1969) and, according to Cheng's hypothesis, the rotation maximum at 293 nm is due to the indole rings being oriented in a certain way. It should be remembered that the spectrum of the intraviral RNA in this case is determined from the difference between rotation of the intact virus and of the repolymerized protein (Cheng, 1968), or of the intact and disrupted virus under conditions preventing renaturation of RNA (Dobrov *et al.*, 1972*a*). It is clear, therefore, that the postulated orientation of the tryptophan residues is absent in the monomeric or repolymerized protein and appears in the nucleocapsid, in all probability, owing to the interaction with the bases of the intraviral RNA. This viewpoint is in good agreement with the above results of the UV dichroism studies concerning the similar orientation angles for tryptophan indole rings and the base planes in the TMV particle (Taniguchi *et al.*, 1971). The possibility of interaction between the bases of the intraviral RNA and the tryptophan of the protein in TMV and PXV is confirmed also by the absence of a peak of positive rotation for particles of CMV and DMV, which are closely related to TMV but contain only one tryptophan residue (Dobrov *et al.*, 1972*a*, 1973). As the subunits of the virus protein have stable secondary and tertiary structures, interaction with the bases may involve only particular tryptophan residues located in the nucleocapsid in the immediate neighborhood of the RNA chain (Fraenkel-Conrat, 1969). According to Cheng's suggestion, these conditions are met in TMV protein by the Trp^{52} residue, but it is quite evident that with small alterations in the secondary and tertiary structure of the protein subunit the interaction of its amino acid residues with RNA may be significantly changed. Unfortunately, there are still no data on the primary structure of the polypeptide chain of CMV and DMV, and nothing definite can be said about the location of the sole tryptophan residue in these proteins.

The above hypothesis is based on the concepts of direct interaction between the bases and the aromatic amino acids which have been developed in a number of studies on model compounds using the methods of nuclear magnetic resonance, fluorometry, CD, and ORD (Gabbay *et al.*, 1972, 1973; Hélène and Dimicoli, 1972; Hélène *et al.*, 1971*a,b*; Buckingham and Danchin, 1973). These authors believe that under certain conditions there may be formed a mixed stack in the form of a sandwich made up of RNA bases and aromatic rings of tryptophan or tyrosine. However, there are certain doubts as to the validity of this concept (Jacobsen and Wang, 1974) and, besides, a simple analogy between the results of the model experiments and the actual interactions between RNA and protein in the rodlike viruses can hardly be treated as proof. In reality, the situation may be even more complicated, as the same authors have reported that the interaction of the tryptophan-containing peptides with polynucleotides either does not cause alteration of the CD spectra or somewhat reduces the amplitude of the positive peak (Gabbay *et al.*, 1972; Hélène *et al.*, 1971*b*). At the same time, the CD spectra of TMV have a form close to that of the RNA spectra in water but with a larger amplitude, similar to the ORD spectra (Miall and Walker, 1968). Hence, we may conclude that either the conditions of interaction between the aromatic amino acids and the bases *in situ* are of different character and may also influence the CD spectra of RNA, or in virions there occur also other interactions between these two partners in nucleoprotein bonds. The experiments with model compounds demonstrate that H bonds may appear between the OH groups of amino acids and the NH_2 groups of the bases (Sellini *et al.*, 1973), and an analysis of the fine chemical structure of the DNA-protein bonds in isometric bacteriophages indicates that protein amino groups interact with the C^4 atom of cytosine (Tikchonenko *et al.*, 1971, 1973). However, there is no direct experimental evidence in favor of such interactions in the rodlike viruses.

As mentioned above, the amino groups of the RNA in TMV do not react with HCHO, even during seven days of incubation, whereas such reactions do occur in CMV, DMV, and PXV (these viruses being in order of decreasing reactivity). These results cannot be ascribed to the difference in permeability of the protein capsid to HCHO (zero for TMV and unlimited for PXV). The radial distribution of the electron density in TMV and CMV is the same with the exception of small portions (see above), and calculations demonstrate that the degree of "porosity" of the capsid in these viruses is sufficiently high and cannot hinder diffusion of low-molecular-weight compounds (Starowsky, 1971), which are known to diffuse easily in protein crystals (Bishop *et al.*, 1966). Therefore, the only reasonable explanation is that the amino group of all three bases in the virus particles studied are involved in interactions with protein, which are strongest for TMV and weakest for PXV. As a result of this, the reaction between HCHO and base amino groups is almost completely inhibited in the former and proceeds relatively rapidly in the latter.

This conclusion is in agreement with the data on the changed reactivity of the RNA bases in TMV to nitrous acid, nitrosoguanidine, and some other chemical modifiers in comparison with the reactivity of the bases in free RNA.

Schuster and Wilhelm (1963) were the first to demonstrate the fact confirmed later by Sehgal (1973), Sehgal and Krause (1968), and Sehgal and Soong (1972) that deamination of free RNA and of RNA in TMV differs significantly. While guanine in the free RNA is deaminated with the highest rate, in the intraviral RNA this base is very unreactive toward nitrous acid. According to Sehgal's data, in 168 hours of incubation at pH 4.0 not more than 21–25% of the available guanine was deaminated in the intraviral RNA, while 70% of the guanine entered the reaction under these conditions in the isometric southern bean mosaic virus. Although none of these authors has reported the rate constants for individual bases, a comparison of their results suggests that deamination of adenosine in TMV as compared with the free RNA is considerably slower, whereas the rates of cytosine modification *in vitro* and *in situ* are approximately the same. Given in Table 4 are the data on the relative rates of deamination of the three bases in various situations (Sehgal and Krause, 1968).

Thus, the rate of deamination of purines in intraviral RNA diminishes in relation to the rate of modification of cytosine. The differentiated changes in the deamination rate for three bases in the viral particle indicate that permeability of the capsid cannot be responsible for the different amino group reactivities. Evidently, such a factor as permeability of the capsid should influence in a similar way the deamination of all three bases *in situ*. Furthermore, the almost complete resistance of the guanine amino group toward the attack by HNO_2 cannot be explained by protection contributed by the intrastacking and complementary interactions since in solution these factors provide only for relatively small deceleration of the reaction rate (Schuster and Wil-

TABLE 4

The Relative Reactivities of Bases[a] toward HNO_2

	Bases		
	Guanine	Adenine	Cytosine
Ribonucleosides	6.5	2.2	1.0
Escherichia coli tRNA	2.1	1.4	1.0
TMV RNA	2.1	1.9	1.0
TMV (intact virion)	0.01	0.53	1.0

[a] The rate of deamination of cytidine is arbitrarily taken to be unity.

helm, 1963). Naturally, these differences between the reactivities of base amino groups *in vitro* and *in situ* lead to biological consequences: the rate of inactivation of infectious TMV RNA is six times that for the intact TMV, and the rate of mutagenesis with respect to the number of surviving infectious centers, as shown in Table 3, is significantly higher in the case of viral particles (Sehgal and Krause, 1968; Singer and Fraenkel-Conrat, 1969*a,b*). The apparent cause of these differences is the fact that formation of xanthine as result of deamination of guanine seems to be a lethal event (Tsugita and Fraenkel-Conrat, 1963; Vielmetter and Schuster, 1960). Evidently, only the interaction with protein may be responsible for such considerable alteration of the reactivity of the base amino groups. The most dramatic change of the guanine amino group reactivity is to a certain extent comparable to the blocking of the free ϵ amino group of lysine in the viral capsid (Perham and Richards, 1968; Tremaine and Goldsack, 1968). Thus, a purine amino group reveals the greatest alteration of its reactivity in the viral nucleocapsid, and this fact is apparently related to the old observation of Fraenkel-Conrat and Singer (1964) that the purine polynucleotides but not the pyrimidine polynucleotides can replace viral RNA in particle reconstitution. Evidently, between purines of the intraviral RNA, especially guanine, and amino acid residues of the protein there is a strong interaction which involves the amino groups of these bases and radically changes their reactivities. Naturally, this does not mean that *in situ* pyrimidines do not take part in the interaction with protein and that, in particular, the amino group of cytosine in the intraviral RNA is free. On the contrary, as evidenced by the stability of the intraviral RNA to hydroxylamine modification [TMV (Schuster and Vielmetter, 1961)] or O-methylhydroxylamine modification [TMV, CMV (A. Mazhul and Dobrov, unpublished)], the C^4 atom of cytosine or its NH_2 group is also involved in the interaction with protein. Of course, the mechanisms of such an interaction may differ for purines and pyrimidines. Though we do not yet know the factors determining the stability of these viruses with respect to hydroxylamine and its derivative, in the phage DNAs the C^4 atom of cytosine is also stable to direct nucleophilic attack (see below), and modification of the intraphage cytosine is effected in an indirect way-by means of attack on the C^5–C^6 double bond, with substitution of the base amino group by the amino group of the nucleophilic amino acid residue (Tikchonenko *et al.*, 1971, 1973). This gives rise to formation of covalent cross-links of DNA, with protein as the intermediate stage of modification. If in TMV the C^5–C^6 double bond is for some reason unavailable for the reaction with O-methylhydroxylamine or if the C^4 atom of cytosine interacts with an amino acid residue which is incapable of the nucleophilic attack, modification by H_2NOCH_3 will not occur.

The composition of the bases in TMV RNA alkylated *in vitro* and *in situ* may yield valuable information on changes in the reactivities of the individual functional groups of the bases. The results of Singer and Fraenkel-Conrat (1969*a,b*) summarized in Table 5 demonstrate that dimethyl sulfate and methylmethanesulfonate *in situ* almost always alkylate only the 7 position of guanine in the intraviral RNA. The reactivity of adenine and cytosine toward alkylation with these agents is greatly reduced, the latter in marked contrast to the results obtained with nitrosoguanidine (see above). Significant alterations in the reactivity of the RNA bases *in situ* can be explained if the 1 position of the purine ring atoms strongly interacts with protein and is, thus, unavailable for other reactions.

Evidently, it is the RNA–protein interaction in the TMV particle that is responsible for the resistance to HCHO of all the bases containing amino groups. Taking into account the availability to formaldehyde of the RNA bases in other viruses studied, it may be of interest to study their behavior in deamination. As for TMV, the results of the biological experiments are in complete agreement with the results of the chemical modification. The isolated TMV RNA has been found to be many times more sensitive to HCHO than the intact virus (Staehelin, 1957, 1958); thus, 0.1% HCHO produces total inactivation of the infective TMV RNA in 60 minutes of incubation while even 24 hours of incubation of TMV with 1% formaldehyde produces only partial inactivation. Elsewhere we noted the difficulties in interpreting these data owing to

TABLE 5

Alkylation of Free and Intraviral TMV RNA[a]

Alkylation products	Intact TMV		TMV RNA	
	DMS	MMS	DMS	MMS
1-Methyladenine	1	0.7	17.5	15.5
3-Methyladenine	0	0	2.7	1.3
7-Methyladenine	2	0.4	3.4	3.5
7-Methylguanine	97	97	66	73
3-Methylcytosine	0.2	1.7	10	6.7

[a] Percentages of the total amount of alkylated bases. DMS, dimethyl sulfate; MMS, methylmethanesulfonate. From Singer and Fraenkel-Conrat (1969*a, b*). Compare with Table 2 for different effect of nitrosoguanidine.

the possible inactivation of TMV due to the reaction between HCHO and the protein component, primarily by formation of cross-links (Tikchonenko, 1969). It has been found that the TMV protein capsid is very peculiar in this respect and that resistance to heating is not increased in the presence of HCHO, in contrast to other viruses (Kust *et al.*, 1972).

To summarize, we may conclude that the amino groups in the intraviral RNA, despite the absence of H pairing, are, nevertheless, not free and their reactivity is altered due to interaction with protein. Undoubtedly, these interactions differ for each of the amino purines and pyrimidines, thus determining the differentiated response of the intraviral RNA to various chemical attacks.

The RNA–protein interactions must not, of course, be thought of as uniform for the whole polynucleotide chain. The sequence of nucleotides in RNA undoubtedly influences the interaction between RNA and protein; this is evidenced by marked differences in the efficiency of heterologous reconstitution of particles, using TMV protein with RNA taken from different sources (Fraenkel-Conrat, 1969). Even though the viral RNA in such experiments may be replaced by purine polynucleotides, we should not forget that in doing so the absolute rate of reconstitution is sharply reduced. In particular, the best results among all the synthetic polynucleotides studied have been obtained with poly (A), but the rate of reconstitution from the TMV protein and poly (A) was 100 times slower than the homologous reconstitution in the TMV-protein–TMV-RNA system.

The close interaction between protein and RNA is shown also by the results of numerous studies of the action of UV light on viruses and, in particular, TMV. Dimers and photohydrates of pyrimidines are known to be the main factors responsible for inactivation of various biological systems irradiated by UV light, in particular, of the infective TMV RNA [see references in Singer, (1971)]. At the same time, intact TMV is inactivated considerably slower than the infective RNA in solution and, as mentioned above, there are no pyrimidine dimers among the photoproducts (Tao *et al.*, 1966, 1969; Singer, 1971; Goddard *et al.*, 1966; Carpenter and Kleczkowski, 1969). Analyses of RNA preparations isolated from the irradiated virus reveal only photohydrates. However, our concepts of the mechanism of TMV inactivation under UV irradiation run the risk of being further complicated since, in addition to the above compounds, two more compounds of unidentified nature were found in RNA extracted from the irradiated virus (Tao *et al.*, 1969).

The absence of dimers in irradiated intraviral RNA is in good agreement with the fact that the irradiated TMV is incapable of photoreactivation, in contrast to the photoreactivation of the irradiated infective TMV RNA (Carpenter and Kleczkowski, 1969; Hurter *et al.*, 1974). It should be recalled here that photoreactivating systems are known to provide for repair of dimerized pyrimidines. It has been demonstrated in experiments with irradiation of TMV RNA in water and in the presence of Mg^{2+} that formation of dimers *in vitro* is enhanced in polynucleotides with ordered secondary structure, primarily owing to ordered base stacking (Singer, 1971; Wang and McLaren, 1972). The special conformation of TMV RNA *in situ* without complementary and interplanar interactions between the bases prohibits or at least hinders formation of dimers. This point of view is supported by the formation of photoreversible, cyclobutane-type pyrimidine dimers in UV-irradiated PXV. Huang and Gordon (1974) have found that 5.7 ± 1.7 dimers of the cyclobutane type are present in each virus particle per lethal biological hit. The number of dimers correlates with the infectivity of the UV-irradiated virus. The presence of pyrimidine dimers induced by UV irradiation agrees with the earlier established ability of PXV to undergo photoreactivation after direct UV treatment (Kleczkowski and Govier, 1969).

At the same time, one must keep in mind that peculiarities of the secondary structure of RNA in rodlike viruses may not be the sole cause preventing the formation of pyrimidine dimers. If this were so, it would be reasonable to expect formation of pyrimidine dimers in UV-irradiated isometric RNA viruses. In virions of this type the secondary structure of the RNA is believed to be the same or nearly the same as *in vitro* and its interaction with the protein is minimum (see below). Nevertheless, Remsen *et al.* (1970) and Yamada *et al.* (1973) reported the absence of UV-induced pyrimidine dimers in R17 and MS2 phages.

Of considerable interest in regard to the structure of the intraviral RNA and its interaction with protein, is the formation of the cross-links induced by UV irradiation in the TMV particles. Strictly speaking, in the case of TMV and other rodlikc viruses no valid evidence of primary cross-links between RNA and protein has been obtained. What Goodard *et al.*, (1966) and Tao *et al.* (1966) have demonstrated is really a much stronger association between RNA and protein in irradiated TMV than in untreated virus. For very high doses the yield of RNA that can be isolated from irradiated virus is definitely lower [C. N. Weber cited by Tao *et al.*, (1969)]. All these data may be well explained by UV-induced cross-linking between RNA and protein. This idea is in good agreement with the results obtained by Smith's group with tissue nucleoproteins and a mixture of amino acids and nucleotides (Smith *et al.*, 1966; Smith and Aplin,, 1966; Smith and Meun, 1968; Smith and O'Leary, 1967). In these experiments, UV irradiation induced protein-nucleic acid cross-linking, and the binding of uracil to amino acids, especially to cysteine, tyrosine and phenylalanine, was directly demonstrated. Such hypothetical RNA–protein cross-links, or, more accurately, labile protein–RNA bonds (Tao *et al.*, 1969), have been used also to explain the fact that irradiated TMV is incapable of reactivation. However, Huang and Gordon (1974) have come to the conclusion that the inability of the irradiated TMV to undergo reactivation is due to the absence of pyrimidine dimers.

The initial scepticism concerning these RNA–protein bonds (Kleczkowski and McLaren, 1967) was based on the fact that deproteinization of the irradiated TMV eliminated protein, but the RNA was, nevertheless, incapable of reactivation, even after reconstitution with nonirradiated TMV protein (Tao *et al.*, 1968). Smith and Meun (1968) seemed to open the way for solving this problem by demonstrating that phenol deproteinization of irradiated TMV indeed destroyed the labile cross-links but simultaneously induced secondary irreparable damage in the RNA. The mechanism of this damage has not yet been established.

The concept of RNA–protein cross-links, in spite of the absence of strong experimental evidence, seems to be very attractive and is also used as an explanation of the results obtained with some isometric viruses. UV irradiation greatly reduces the injection of RNA by MS2 (Yamada *et al.* 1973) or f2 phage (Werbin *et al.* 1968) into the infected cells, the adsorption of the irradiated phage being unchanged.

The problem of RNA–protein cross-linking should be considered in more detail in connection with chemical modification of viruses by *O*-methylhydroxylamine and bisulfite, but here we shall restrict ourselves to indicating that formation of cross-links should be associated with the preexisting RNA–protein interaction.

The reduced sensitivity of the double bond of the pyrimidine rings to photohydration (Tao *et al.*, 1966) also seems to be associated with the peculiar conformation of the intraviral RNA. These authors have found that free TMV RNA is inactivated by UV light in H_2O faster than in D_2O, whereas the intact virus has the same rate of inactivation in both cases. This fact is most probably due to the low level of photolysis of the double bonds in RNA pyrimidine residues *in situ* which may take place if these parts of the pyrimidine rings are in a hydrophobic environment (Tao *et al.*, 1969). This assumption is contrary to the views of some authors (Tremaine and Goldsack, 1968; Perham and Richards, 1968) who believe that the RNA should be in a hydrophilic environment in TMV particles. However, this inconsistency is eliminated if we proceed from the concept of microheterogeneity of the RNA structure *in situ,* meaning, in particular, that different components of the polynucleotide chain may have different environments.

The above-described differences in the conformation of RNA in TMV and PVX particles correlate well with a deuterium isotope effect observed in irradiated PVX preparations by McLeary and Gordon (1973). This effect evidences the easy permeability of the inner parts of PVX virions for the solvent. At the same time, individual properties of different RNAs may greatly influence their deuterium isotope effect; for example, its value for free TMV RNA is 2.0 (Tao *et al.*, 1966), in contrast to 1.2 for free Qβ phage RNA (Furuse and Watanabe, 1971). Due to this difference, one must be careful in calculating the deuterium isotope effect value, especially when different RNAs and different viruses are compared.

It is important to note that in TMV and its various strains greatly different sensitivities toward UV light have been demonstrated, whereas the radiosensitivities of their isolated RNAs are approximately equal (Bawden and Kleczkowski, 1959*a*; Reichmann, 1960). In particular, the inactivation curves for PXV are much steeper than those for TMV, which is in agreement with other facts indicating the weaker RNA–protein interaction in the potato virus. The experiments with TMV strains U_1 and U_2 have demonstrated that the differences in the UV sensitivities of these strains may be related to the fine peculiarities of their structural organization (Siegel *et al.*, 1956; Siegel and Norman, 1958). In particular, the strain U_2 is 5.5 times more sensitive than the strain U_1 to irradiation in the moist state, whereas the sensitivities of the dried preparations to irradiation are the same and do not differ from the sensitivity of the infectious RNAs. It should be stressed that the sensitivities of the dried and moist preparations of the strain U_1 are the same and approximately equal to the sensitivity of the free RNA. Evidently, drying disturbs the RNA–protein interaction in U_1 particles and concurrently deprives the protein of this strain of the ability to protect the RNA against UV damage. Upon irradiation of the hybrid particles reconstituted from RNA and protein of various strains, the sensitivity to UV light is typical of the strain that donated the protein (Rushizky *et al.*, 1960; Streeter and Gordon, 1968). The radioprotective action of viral protein was observed only for the wavelengths corresponding to the absorption maximum of nucleic acid and was ascribed to the migration of energy from RNA onto protein. If the irradiation wavelength is 230 nm, an enhancing effect of the protein instead of protection is observed (Goddard *et al.*, 1966). It is suggested that the reverse effect takes place in this case–the energy absorbed by the virus protein migrates onto RNA, although the mechanism of this effect and the pathways of energy migration are still unknown for both cases.

So far, we have discussed the direct interaction between the RNA bases and the capsid amino acid residues. However, in the viral particle there should exist such traditional nucleoprotein bonds as the salt interactions between RNA phosphates and positively charged amino acids, even though there is no direct experimental evidence of this. The old results of Kausche and Hahn (1948) are doubtful in view of insufficient amount of controls. More recent studies in this field by Fraenkel-Conrat and Colloms (1967) have yielded more definite though by no means unambiguous results. These authors made use of selective acetylation of lysine ϵ amino groups together with the most reactive tyrosines. In the isolated virus protein, both the Lys^{53} and Lys^{68} residues were modified, whereas in the intact virus only Lys^{68} residue reacted. The amino group of the Lys^{53} residue may be in ionic linkage with a negatively charged group of the virion, which may be an RNA phosphate group or a carboxyl of the same or a neighboring polypeptide chain. The possibility of RNA involvement appears to be supported by the finding that reconstitution of virus from a mixture of protein and RNA is impossible after prior acetylation of the Lys^{53} residue. Modification of the Lys^{68} residue under similar conditions does not influence the reconstitution. Proceeding from these results, it may be suggested that the phosphate groups of one third of all nucleotides (see above) are involved in salt interaction with a lysine residue. The state of the remaining two thirds of the phosphate groups of the intraviral RNA is so far unknown, but very probably they are bound to arginine residues.

1.2. Filamentous DNA-Containing Phages

An essentially different structural organization is revealed by the single-stranded ring DNA in the particles of filamentous bacteriophages, one of which, the phage fd, has been studied by a number of authors. Unfortunately, the information about this phage and its nucleic acid is extremely scarce as compared to its phytopathogenic relatives discussed above, hindering considerably a comparative analysis and interpretation of the differences observed.

The electron microscopy and X-ray diffraction results show that the capsid of the phages of the genus *Inovirus* is a helically organized tube (Marvin, 1966; Marvin and Schaller, 1966; Bendet and Mayfield, 1967; Rossomando and Zinder, 1968; Rossomando and Milstein, 1971; Wiseman *et al*., 1972). The mean length of the virions for the phage fd group is 870 ±30 nm, and the mean length of the phages in the If group is 1300 nm. The most probable diameter of the virions is 5.5 nm although diameters reported for various specimens of this group range from 5 to 6.8 nm [see reference in Marvin and Hohn (1969)]. Concentrated preparations of the filamentous phages yield paracrystals which may be oriented and studied using the X-ray diffraction techniques (Marvin, 1966).

The best studied specimen of the genus *Inovirus* is the phage fd; its capsid is constructed of 300 protein subunits of molecular mass 5.17×10^3 daltons, and one to three molecules of A protein (Marvin and Hohn, 1969). The pitch of the protein helix is 1.6 nm and there are 5.4 protein subunits per turn of the helix. Two models of the structure of the phage fd and related viruses have been suggested. According to one (one-strand model), the virion is a cylinder with a central cavity along the long axis of the particle where the DNA is located (Bradley, 1967). In the second model, the so-called two-stranded model, the virion is made up of two protein cylinders surrounding the DNA strand (Marvin, 1966; Takeya and Amako, 1966; Minamishima *et al*., 1968; Day, 1969; Marvin and Hohn, 1969; Eiserling and Dickson, 1972; Stroke and Haliova, 1972; Griffith and Kornberg, 1972; Frank and Day, 1970; Marvin *et al*., 1974). Two subclasses of the two-strand model are possible: either the axes of the two helical strands could be parallel to each other, or each of the two helical nucleoprotein strands could be itself wrapped around another axis (coiled coil).

It is not necessary for the DNA strand in these models to follow the arrangement of protein subunits. Both models are compatible with available experimental data and no choice can be made between them, although the two-strand model is more popular.

However, a detailed X-ray diffraction analysis of this phage is hindered by the lack of success in repolymerizing the protein in the form of empty capsids devoid of DNA. Therefore, nothing is known about the tertiary structure of the intraviral DNA and its relations with the capsid protein helix. According to Marvin (1966) and Day (1969), if the phage particle length is about 870 ± 20 nm and the contour length of the DNA ring molecule is about 1600 nm, corresponding to the molecular mass of 1.9×10^6 daltons, the DNA molecule may be placed along the virion in the form of two polynucleotide strands following the shape of the protein helix, provided that the distance between nucleotides (with respect to the particle axis) is not more than 0.27 ± 0.02 nm. In this case there are 12 nucleotides per turn of the helix. The above distance between nucleotides in the phage DNA *in situ* proves to be the same as in the A configuration of double-stranded DNA, which is known to exist only at a relative humidity of about 70% (Langridge *et al.*, 1957). As there are grounds to believe (see below) that the angle of inclination of the planes of the bases to the particle axis (20–30°) is similar to that in the A configuration of DNA, it is quite possible that the packing of the single-stranded DNA in the phage fd is even more similar to the geometry of the polynucleotide chain in the double-stranded complex of the A configuration, although the ORD and CD spectra displayed by fd phage suspensions do not support this interesting hypothesis.

As shown by the results of UV spectrophotometry (Day, 1969; Rossomando and Zinder, 1968; Rossomando and Bladen, 1969; Wiseman *et al.*, 1972), the DNA in the virions of all examined viruses of this group shows highly ordered base stacking since the calculated value of ϵ (P) at 260 nm is 6750 ± 300, as compared to 6950 for free DNA in 0.15 M Na^+ and 8280 in 0.0015 M Na^+ (results for the phage fd). These values clearly suggest that under favorable conditions *in vitro* and in virions the DNA of the phage fd shows the maximum hypochromism characteristic of absorption by native double-stranded DNA in solution (Inman and Jordan, 1960). Therefore, disruption of the phage and release of the DNA into solution at room temperature and physiological ionic strength are not accompanied by a change in absorption at 260 nm, whereas under conditions causing denaturation of DNA there is observed a hyperchromic effect. The hypochromic state of the phage DNA *in situ* cannot be ascribed to formation of a perfect double helix since in this DNA the rule of complementarity is not satisfied and the ratio A/T = 0.66 and G/C = 0.91 (Hoffman-Berling *et al.*, 1963). The low specific absorption of phage-fd DNA is apparently associated with the perfect base stacking in the ring molecule, which equally determines the A_{260} value both *in situ* and *in vitro*. However, it is clear that even if under these circumstances there occur interactions with protein, this does not lead to disturbance of the intrastacking interaction of the type described above for the plant viruses.

A very interesting feature is observed in the UV absorption spectra of the phage fd in the 270–300 nm range (Day, 1969); compared to the absorption of the free DNA, at this range of the spectrum the phage shows a hyperchromic state which disappears with the destruction of the phage. As a result, DNA release from the phage into solution is accompanied by a hypochromic effect in the long-wavelength part of the spectrum. A similar situation is observed upon disruption of large isometric phages containing double-stranded DNA (Tikchonenko *et al.*,

1966), where hypochromism in the 270–300 nm range is 2–2.5 times greater than hypochromism at 260 nm (Tikchonenko *et al.*, 1974*a*). At present this may be explained by anomalous scattering in the 270–300 nm region of the absorption spectrum or by the conformational changes of phage components. The first possibility will be considered later in connection with UV absorption spectra of large isometric phages, but it is worth mentioning that no anomalies in UV absorption spectra have been found in RNA-containing filamentous and isometric viruses (see above and below).

The available data do not allow one to determine which phage component (DNA or protein, or both) is responsible for the increased absorption in this spectral range. The analogy with the large phages, despite the similarity of effects, may prove to be a false one as these phages are distinguished by a high percentage of DNA (40–50%, compared to 12% for the phage fd). Therefore, while the protein contribution to absorption at these wavelengths is small in phages of the T2 or S_d types (Tikchonenko *et al.*, 1966), in the phage fd it is rather significant. Finally, under mild conditions disruption of virions of the large isometric phages preserves the protein component in the form of empty capsids, whereas such capsids cannot be obtained from the filamentous phages, either by destruction of virus or by repolymerization of protein. As the protein absorption may differ considerably for the monomeric and polymeric forms, the phage fd experiments meet with very complicated problems of getting adequate controls to allow for absorption by the protein component. However, even if hyperchromism in the long-wavelength part of the spectrum is due to DNA, similar to the large isometric phages, the interpretation of the results is very difficult.

In principle, two groups of factors may be responsible for hyperchromism of polynucleotides in the 270–300 nm range: (1) alteration of the state of the chromophore groups due to the external effects (protonation, hydrogen bonding, variation of the refractive index, etc. [see reference in Wetlaufer (1962) and Schellman and Schellman (1966)]; and (2) alteration of the electron interaction system in the chromophores in the range of $n \rightarrow \pi^*$ and $\pi \rightarrow \pi^*$ transitions owing to the different chain geometry and arrangement of the base planes [see Tinoco (1960)]. If the hyperchromic state at 270–300 nm of both large isometric phages with double-stranded DNA and small filamentous phages with single-stranded DNA have the same causes, the first group of factors seems to be responsible. It may be easily appreciated that the orientation effects capable of influencing the interaction of chromophores should be considerably stronger in asymmetric filamentous viruses than in isometric viruses, where DNA segments do not have a preferential orientation with respect to the particle axis (Tikchonenko, 1969).

A number of authors have determined the orientation of the bases of DNA of phage fd with respect to the long axis of the filamentous virion using the techniques of UV electric or flow dichroism (Bendet and Mayfield, 1967; Rossomando and Milstein, 1971; Cram and Deering, 1970). All these authors agree that, in contrast to TMV, the chromophores in the fd virions are more or less perpendicular to the long axis and the angle of inclination of the bases is 20–30°.

The orientation of the bases of the intraphage DNA may be changed by heating the phage suspension to a subcritical temperature at which the virion is not destroyed. Under these conditions the inclination angle of the DNA bases in the phage f1 is increased by 1.5°, while optical properties are not changed (Milstein and Rossomando, 1972); this effect has been ascribed to variations in the DNA–protein interactions.

The interpretation of the above results is hindered by the inability to obtain empty capsids of the phage fd, in addition to limitations of the techniques of birefringence and UV dichroism (see above). According to Bendet and Mayfield (1967), the tryptophan and possibly the tyrosine in the phage fd are oriented practically parallel to the long axis of the particle, similar to TMV. In other words, the planes of the chromophores of the aromatic amino acids and the planes of the DNA bases have different orientations in the phage fd, making mixed base–tryptophan stacking impossible. It may actually be this factor that is responsible for the absence of any anomalies in the ORD spectra of these phages.

The ORD spectra (Day, 1966) do not yield any additional information on the organization of the intraviral DNA, partly, due to the fact that nobody has managed to obtain the empty protein capsids of this virus. This prevents the construction of a more or less correct differential ORD spectrum for the DNA *in situ*. In contrast to TMV, phage fd does not reveal positive rotation maximum in the 260–300 nm range. Free phage-fd DNA is distinguished by an extremely small positive rotation with a wide maximum between 280 and 290 nm.

The studies of natural and artificial complexes of the phage-specific protein G5 with the phage DNA have yielded extremely interesting results, revealing to a certain extent the structure of the DNA in the filamentous phages and its interaction with protein (Day, 1973; Alberts *et al.*, 1972). This protein is a product of the gene *5* and a precursor of the mature virus particles which consist of the G8 protein. The complexes of fd DNA with protein G5 are, as the phage itself, filamentous in shape and longer than the mature virions by 20%. In such DNP there are four nucleotides per protein subunit. Despite the external similarity between the mature virions and viruslike DNP particles, their optical properties differ sharply. In particular, the DNA complexed with the G5 protein is in a state close to maximum hyperchromism [$\epsilon(P)$ = 8080, as compared to 6950 in the phage fd]. The measured value of $\epsilon(P)$ does not depend on the ionic strength of the medium, indicating stabilization of the hyperchromic state due to the interaction with protein. The picture observed indicates an almost complete disturbance of intrastacking and complementary interactions in the DNA molecule *in situ*.

At the same time, such structures reveal considerable alteration of the CD spectra, which is similar to that described above for other filamentous and rodlike viruses: a sharp decrease of the amplitude of the positive CD band and its shift to the long-wavelength region of the spectrum. It is suggested that the alteration of the optical properties of the DNA is due to the interaction of the DNA with protein, which is determined primarily by its tyrosine residues. The protein G5 contains five tyrosine residues, determining the strong CD band at 228 nm (there is no tryptophan in this protein). Participation of tyrosine in various interactions, including that with DNA, is accompanied by variation of CD in this spectral range (Friedman and Ts'o, 1971; Adler *et al.*, 1972; Hélène, 1971). It is the complex of the protein G5 with the single-stranded DNA that demonstrates the characteristic decrease of the amplitude of this band and its shift to the long-wavelength region of the spectrum.

The above variation of the optical properties of DNA is specific for the interaction with the G5 protein and is absent in the interaction with the capsid G8 protein; this seems to be due to the profound differences in the secondary structure of these proteins. The G5 protein is almost devoid of a α helix, whereas the G8 protein is distinguished by a high percentage of α spiralization (Alberts *et al.*, 1972). The secondary structure of DNA is also important for the interactions of this type. In particular, the G5 protein does not influence the optical properties of the double-stranded ring DNA.

1.3. Isometric RNA-Containing Viruses

For the last few years, excited discussions have been going on in the virological literature concerning the interaction forces between various viral components and their subunits and the contributions of various types of interactions to the total physical stability of the virion. Even the first attempts to draw some general conclusions from the available experimental data (Kaper, 1972, 1973) have shown that all the viruses, including the isometric ones, are distinguished by great structural diversity. According to Kaper, we may distinguish two borderline virus groups—one with prevailing protein–protein interactions, and the other with prevailing RNA–protein interactions. For viruses of the first group the main factor in the physical stability of the virion is the hydrophobic interaction between the protein subunits in the capsid; the contribution of the RNA–protein interactions to the total stability of the virion is small. The prototype of this group is the turnip yellow mosaic virus (TYMV), and this group includes the wild cucumber mosaic virus, the poliovirus, the phage f2, and many others. For the viruses of the second group the main factor in virion stability is assumed to be the interaction between the protein subunits and RNA. The prototype of the viruses of this group is the cucumber mosaic virus (CMV), and this group includes the broad bean mottle virus, the brome mosaic virus, and others. The main difference between these two groups of viruses is that the protein capsid of the viruses of the first group can exist in solution in the form of empty protein shells due to the strong interaction between the protein subunits; the capsid may also be obtained as a result of self-assembly of the protein subunits in the absence of RNA in the medium. In the first-group viruses the RNA–protein bonds exist only within the weakly acidic *p*H range and are easily disturbed, even at neutral *p*H (Kaper, 1971, 1973; Jonard, 1972). This feature of TYMV and other viruses of the first group has been extensively studied in experiments concerning RNA degradation *in situ* under various conditions, and we shall discuss it in detail below.

For the viruses of the second group, owing to weaker bonds between the protein subunits, empty capsids are either unobtainable, or the capsids or their large fragments degrade rapidly when attempts are made to separate the shell from RNA (Kaper, 1971, 1972, 1973). It is this feature of CMV which is responsible for the characteristic degradation at high salt concentrations, causing rupture of the RNA–protein bonds (Kaper, 1971, 1973; Kaper and Geelen, 1971). As for the viruses of the first group, the stability of their virions, as a rule, increases in media with high salt concentrations. Judging by the sensitivity of CMV and other viruses of this group to various external factors, they are similar to polyelectrolytic complexes existing due to the strong salt bonds (Kaper, 1971). Naturally, CMV and similar viruses are not capable of self-assembly of the protein subunits in the absence of RNA (van Kammen, 1972; Bancroft, 1970).

A useful criterion for classification of viruses according to the prevailing type of interaction is their sensitivity to low concentrations of ionic detergents (e.g., sodium dodecyl sulfate), which destroy viruses of the second group but do not influence the viruses of the first group (Kaper, 1971, 1973). It should be stressed that between these two extreme groups there are numerous intermediate and transitional forms which combine in various proportions the properties of the first and second types of viruses. At the same time, one must not think that in the viruses with strong RNA–protein interactions the structure of the RNA *in situ* should be altered significantly or that in viruses with prevailing protein–protein interactions the opposite effects should be observed. The real situation is much more complicated. The classification of viruses into groups with prevailing protein–protein interaction or with prevailing RNA–protein interaction takes into account, primarily, the relative contribution of these factors to the physical stability of the virion and its capsid. The aspects of the nucleoprotein interactions and any resulting alterations of RNA conformation *in situ* are usually ignored. For instance, the salt interactions between protein and nucleic acid are known to affect only slightly the secondary structure of the latter. If the salt bonds prevail in the RNA–protein interaction it may be presumed that the RNA structure *in situ* will be changed insignificantly. Similarly, the interaction with protein may involve primarily single-stranded portions of the RNA so that the secondary structure of the intraviral nucleic acid may be little affected. However, we cannot rule out other interactions between RNA and protein which may involve the helical portions of the RNA. In this case we could expect more or less significant conformational alterations of the polynucleotide chain. Also, we should not forget about significant interspecies differences in the structure of the intravirion RNA for even closely related viruses. The structural peculiarities of the virions and the predominant interactions in the nucleocapsid should apparently influence the mode of formation of the viral particles in the infected cells. Although our knowledge of this problem is clearly inadequate, it may be expected that for the viruses of the first group capsid formation and virion maturation may be separated in time. Indeed, during poliovirus synthesis the protein capsid is formed first and then it is combined with RNA, thus giving rise to mature virions. In other words, the capsid is the precursor of the virion; such formations in the infected cells are called procapsids (Jacobsen and Baltimore, 1968). Interestingly, the compositions and properties of the procapsids and the mature capsids are not identical. Of the four polypeptides found in the mature virion (VP1, VP2, VP3, and VP4), the procapsids contain only VP1 and VP3, as well as the not-yet-cleaved protopeptide VP0, which yields VP2 and VP4 in the mature virions (Jacobsen and Baltimore, 1968).

It is quite evident that notwithstanding the importance of a virion's physical stability, this stability cannot be the primary evolutionary goal for viruses in general. Otherwise, the evolutionary ideal for viruses would be to form some universal "living safes." The interaction of nucleic acid with protein and the alteration of RNA structure *in situ* should be considered not only from the viewpoint of their contribution to the physical stability of the particle, but also in terms of other biological goals (e.g., primitive differentiation, effective adsorption, easy release of nucleic acid from the capsid, possibility of various RNA–protein recognitions, etc.). The result is the biochemical and biophysical manifold of structural organization which provides for filling various biologic niches. Therefore, it may be easily conceived that there exist in viruses of the first group, along with strong protein–protein interactions, some considerable RNA–protein interactions including not only salt bonds but also interactions involving the rings of the bases and the aromatic amino acids.

One of the methods of revealing interactions involving the bases in RNA is the determination of the reactivities of various functional groups of the bases. One highly promising approach deals with transformation of the preexisting secondary interaction forces into covalent bonds between DNA and protein; this approach was first introduced for large phages with double-stranded DNA (Tikchonenko *et al.*, 1971, 1973). When we expose a virus to hydroxylamine and its derivatives, to bisulfite (Skladneva *et al.*, 1973), or to UV light (Simukava and Budowsky, 1974), the C^5–C^6 double bond of the pyrimidines is saturated, leading to activation of the C^4 atom. Under the conditions of close interaction between bases and protein, the C^4 atom may react with nucleophilic amino acids. Following these events, the amino group at the C^4 atom of the cytosine residues may be substituted by the amino group of that nucleophilic amino acid with which it interacted *in situ*. In other words, at the site of the preexisting interaction between base and protein a covalent cross-link appears. In particular, Turchinsky *et al.* (1974) have demonstrated that when the phage MS2 is treated with bisulfite, four to five protein subunits are cross-linked to RNA with covalent bonds, including the sole A-protein molecule in the virion.

Undoubtedly, the isometric viruses should differ from the rodlike viruses with helical symmetry in the extent of the influence exerted by the protein component on the structural characteristics of the RNA *in situ*. These differences are primarily determined by the fact that in the isometric viruses the amount of protein per unit weight of nucleic acid is considerably less. As a result, their capsid is, in the final analysis, a relatively thin-shell surrounding nucleic acid. It is clear, *a priori*, that in virions of this type the contacts between the RNA and the protein are limited and a considerable part of the nucleic acid has to keep its own company. The extreme form of this viewpoint amounts to the known concept of the "central body" (nucleic acid) which no more interacts with its outer shell than "a hat in a hat box interacts with it." The early results of X-ray diffraction analysis of isometric viruses have been interpreted along the lines of this concept (Katz and Rich, 1966; Fishbach *et al.*, 1965). This concept of the tertiary structure of RNA is in complete agreement with the interpretation of the experimental data on the secondary structure of RNA in the isometric viruses, which amounts to the conclusion that the RNA structures *in situ* and *in vitro* are the same.

In reality, the tertiary structure of the RNA in viruses of this type cannot be represented in the simplified form of a centrally located spheroid. As shown by the results of later studies, the tertiary structure of the RNA *in situ* is quite complex and is determined to a great extent by the organization of the protein component, even though the nucleic acid–protein contacts here are considerably more limited than in the rodlike viruses. Clearly, the interaction with protein cannot fail to influence the conformation of the RNA, i.e., to influence the characteristics of its secondary and tertiary structures, even though in comparison to the TMV-like viruses the degree and forms of this influence are obviously weaker. Even if in RNA *in situ* there occurs no radical change in the percentage of spiralization and no significant variation in the proportions of single-stranded and double-stranded regions, other alterations of conformation may take place, additional stabilization may appear, changes in the chemical reactivity of the functional groups may occur, etc.

1.3.1. The Tertiary Structure of RNA

The fathers of the new concept of the tertiary structure of RNA in isometric viruses were the biophysicists from Cambridge studying the plant viruses (Klug and Finch, 1960; Klug *et al.*, 1961, 1966*a,b*; Finch *et al.*, 1967*a,b*; Finch and Klug, 1967) and, later, Zipper *et al.* (1971), who studied the RNA-containing phages. The first group of workers studied mainly the X-ray diffraction of crystalline virus preparations (wide-angle scattering) and utilized electron microscopy, while the second group studied small-angle X-ray scattering in virus suspensions.

The principal general conclusion from the studies of the tertiary structure of RNA in isometric viruses of plants and bacteria is that in virions there is no distinct boundary separating the protein shell from the central region occupied by RNA. The data obtained for the phages fr and R17 are summarized in Table 6.

Comparing the data in Table 6, we see that not more than 80% of the virus RNA is concentrated in the "nucleus." The remaining part of RNA is localized at the distance from 10.7–10.8 to 10.4–10.5 nm from the center, forming a mixed RNA–protein zone where the close interactions between these components are feasible. This zone is distinguished by excessive electron density due to the concurrent presence of RNA and protein. Similar effects have been observed earlier in phage preparations by Fishbach *et al.* (1965), but these authors proceeded from more conservative concepts in interpreting their results, assuming that the zones of RNA and protein do not overlap at all. The available techniques of the small-angle X-ray scattering cannot yield sufficient data on the RNA structure in this mixed region. Additional data on this problem have been obtained using the models of the plant viruses TYMV and broad bean mottle virus by means of wide-angle X-ray diffraction. The results for both these viruses have proved to be identical, even though according to Kaper's classification (Kaper and Geelen, 1971) the first virus belongs to the group of viruses stabilized by protein–protein interactions and in the virions of the second virus the prevailing interaction is between RNA and protein. Despite these differences, the tertiary structure of the RNA *in situ* is the same in

TABLE 6

Dimensions in the Intercalation Model of the Phages[a]

Component	fr	R17
RNA		
Mean internal radius	1.0 nm	1.5 nm
Mean external radius	10.74 nm	10.82 nm
Mean excess of electron density	0.0047e nm^{-3}	0.0046e nm^{-3}
Capsid		
Mean internal radius	10.48 nm	10.51 nm
Mean external radius	13.14 nm	13.2 nm
Mean excess of electron density	0.0067e nm^{-3}	0.0066e nm^{-3}
Radius of gyration	10.52 nm	10.57 nm
Volume	0.95×10^6 nm^3	0.963×10^6 nm^3

[a] From Zipper *et al.* (1971).

TYMV and in broad bean mottle virus and is due to close interaction with protein. These results, which have been discussed in detail earlier (Tikchonenko, 1969), may be summarized as follows: the X-ray scattering curves for the intact virus are not simple sums of diffractions due to the mixture of the virus RNA and the empty protein capsids ("the top component"), and the X-ray diffraction patterns due to the protein component of the intact virus and to the empty capsids are identical within the errors of measurement (Klug *et al.*, 1966*a,b*; Finch *et al.*, 1967*a*; Finch and Klug, 1967). The isomorphic structure of the protein in nucleocapsids and isolated capsids permits the derivation of the differential diffraction curves for the intraviral RNA by the subtraction of the diffraction of protein from the diffraction of intact virus.

The TYMV and broad bean mottle virus and their capsids are known to have the shape of an icosahedron with a radius of 14–15 nm and are formed of 180 subunits (polypeptide chains) organized into 32 capsomers, or morphological subunits (12 pentamers and 20 hexamers). The capsomers are seen only in electron micrographs of intact virus particles and are indistinguishable in empty capsids. A comparative analysis of the X-ray and electron microscopy data shows that the RNA of the virus particle has the internal organization corresponding to capsomers and forms 32 local condensates (bumps) with a mean radius of 12.5 nm. Transversely, these bumps embrace groups of 5 (pentamers) or 6 (hexamers) protein subunits, so that the distance between the bumps is 8.5 nm. Thus, all these data indicate that the organization of RNA in these two isometric viruses *in situ* is characterized by the icosahedral symmetry determined by the capsid symmetry. Unfortunately, due to the lack of crystalline preparations similar studies of the RNA-containing phages are unfeasible.

The RNA in the virions of the broad bean mottle virus has a similar structural organization. The small-angle X-ray scattering results indicate that the RNA of this virus forms two zones *in situ* (Finch and Klug, 1966; White and Fischbach, 1973). One zone is located 9.0–13.0 nm from the virion center (the virion diameter is 14.7 nm) and is represented by part of the polynucleotide chain closely interacting with protein. Owing to this interaction, this region of the virion is distinguished by a comparatively poor permeability for sucrose. The remaining part of the RNA molecule occupies the zone nearer to the center. One of its boundaries is the central cavity, the radius of which in the broad bean mottle virus is 3.0 nm, and the second boundary is the interface with the zone of RNA–protein contacts (9.0 nm from the center). A high degree of porosity determined by sucrose permeability in conjunction with high electron density indicate that the RNA strand, similar to that of TYMV, forms bumps in the zone of RNA–protein contacts.

Apparently, the same tertiary structure may be ascribed to the RNA in the virions of some other isometric viruses, though there is no direct X-ray diffraction evidence for this. Such a possibility is suggested by selective degradation of the RNA *in situ* into fragments of a size corresponding to approximately $\frac{1}{32}$ of its molecular weight. These results are discussed separately in the following section.

The second peculiarity of the structural organization of the RNA in isometric viruses is the existence of the central cavity in virions, confirmed by the results of the small-angle X-ray study. This cavity has been shown to exist in a number of plant and bacterial viruses: wild cucumber mosaic, broad bean mottle, brom mosaic, RNA-containing phages R17 and f2, as well as the phage ϕX174 containing single-stranded DNA (Anderegg *et al.*, 1963; Finch and Klug, 1967; Finch *et al.*, 1967*a,b*; Fischbach *et al.*, 1965; Hiebert *et al.*, 1968; Katz and Rich, 1966; Zipper *et al.*, 1971; White and Fischbach, 1973). There are no data about other viruses and, in particular, Klug *et al.* (1966*b*) stress that what is claimed for TYMV is not the absence of a central cavity in general, but the absence of a cavity with a diameter exceeding 7.0 nm, since structures of smaller diameter cannot be found with the available techniques. It is suggested that the smaller diameter of the central cavity in TYMV may be due to the higher relative content of RNA, which amounts to 37% as compared to 22% in broad bean mottle virus (Hiebert *et al.*, 1968). For the plant viruses, the reported cavity diameters range from 6.0 to 12.0 nm, whereas in bacteriophages they are considerably smaller, 1.0–3.0 nm as reported by various authors.

The causes for these differences in the central-cavity size between various viruses are not clear. On one hand, they may be due to the actual differences in the virion structure, as well as to the different percentages of RNA. On the other hand, they may be due to experimental and calculation errors which occur when determining the dimensions of such small structural regions. Thus, in the broad bean mottle virus the central-cavity diameter is estimated to be 6.0–11.0 nm (Finch and Klug, 1967; White and Fischbach, 1973).

If we remember that a central cavity also exists in large bacteriophages containing double-stranded DNA [see reference in Tikchonenko (1969)], it may be concluded that the existence of this structure should be associated with some general principle of organization of isometric virions and the packing of their nucleic acids.

There is no consensus about the reasons for the existence of a central cavity. Klug *et al.* (1966*b*) and Finch and Klug (1966) believe that the major mass of the RNA in the form of radial strands is concentrated in the proximity of the protein subunits. This conclusion is based on the differences between the curves of the small-angle scattering for the TYMV crystals in 1 and 4 M ammonium sulfate. The reflections ($4n + 2$) typical for this type of packing and seen in 1 M ammonium sulfate solution disappear when the salt concentration is increased to 4 M. Similarly, under these conditions the effective radius of the RNA decreases from 12.5 to 11.7 nm. The authors consider it to be most probable (although there may be another explanation) that salt bonds which dissociate in the 4 M salt exist between the radial RNA strands and the protein subunits. Under normal conditions these salt bonds keep the RNA strands in contact with the protein subunits, providing for the appropriate orientation of the RNA in the virion. In this case, the cavity is the virion's central region where there are no RNA strands contacting the protein. This hypothesis does not satisfactorily explain the complete reversibility of the action of 4 M ammonium sulfate; after the elimination of the ammonium sulfate the initial effective radius of the RNA and the reflections ($4n + 2$) are restored. Unfortunately, there is still no confirmation of this interesting observation.

Anderegg *et al.* (1963) suggest that the central cavity in various viruses is due to the force of electrostatic repulsion between the segments of RNA. This concept is based on the assumption that the negatively charged phosphates are poorly shielded and do not interact with protein or metal ions. A significant electrostatic repulsion is possible only with loose packing of RNA in the internal regions and a high degree of hydration, as is the case in reality. Table 7 summarizes the respective data (Zipper *et al.*, 1971) for the RNA-containing phages. These authors have estimated that RNA occupies no more than 18% of the available volume of the virion.

It has already been mentioned that in spite of the limited number of viruses whose structure has been studied by X-ray diffraction analysis, the available experimental data support the concept of the "icosahedral" RNA packing for a large number of viruses. This extrapolation is based on the data on the selective degradation of RNA in some plant viruses.

Kaper and his co-workers (Kaper, 1968; Kaper and Jenifer, 1965, 1967, 1968; Kaper and Halperin, 1965) and some other authors (Matthews and Ralph, 1966; Stols and Veldstra, 1965; Bosch *et al.*, 1967) have demonstrated the possibility of limited degradation of the RNA of TYMV *in situ* under certain conditions. This effect is brought about by various types of treatment: short exposures to alkali, heating, incubation with *p*-chloromercuribenzoate or alcohol, etc. Among these studies

TABLE 7

The Degree of Swelling and Hydration of the Phages f2 and R17[a]

Components	Parameters	f2	R17
Intact phage	Mean degree of swelling	2.35	2.38
	Hydration (g/g)	0.91	0.93
RNA "nucleus"	Mean degree of swelling	5.32	5.45
	Hydration	2.30	2.35
Protein capsid	Mean degree of swelling	1.52	1.55
	Hydration	0.38	0.40

[a] The mean degree of swelling is the ratio of the volume of dissolved phages to the dry volume.

we may class the experiments on cowpea chlorotic mottle (CCMV) and related viruses (Bancroft *et al.*, 1968*a*; Kaper, 1971, 1973), the RNA of which was subjected to limited degradation *in situ* at *p*H valves of 5–7. The viruses remained intact under this treatment. All these viruses are characterized by prevailing protein–protein interaction and relatively weak bonds between RNA and protein which break upon neutralization (see above). However, the direct cause of RNA degradation *in situ* is still a matter of discussion. Some authors believe that degradation is caused by RNase penetrating the capsid, which has been altered by the treatment. Other authors think that phosphodiester bonds are broken due to stresses in the RNA chain caused by changes in conformation of the protein subunits. Furthermore, when alkali is used in such experiments it may serve as the cause of the observed RNA degradation. However, the principal feature of this degradation is its limited and selective character, irrespective of the direct cause of the rupture of the phosphodiester skeleton of the molecule.

If we subject the virions to phenol deproteinization after one of the above types of treatment, we obtain low-molecular-weight but, nevertheless, monodisperse RNA preparations. For the alkaline treatment of TYMV, the sedimentation constant of the RNA was 5 S and the molecular weight of the fragments obtained was $57 \pm 3.5 \times 10^3$. Similar RNA degradation has been observed for CCMV when it is converted from the compact form with the sedimentation constant of 88 S at *p*H 5 to the swollen form with the sedimentation constant of 72 *S* at *p*H 7. Although these forms may convert into each other the, transition from *p*H 5 to *p*H 7 proves to be lethal for the virus since it is associated with loss of infectivity. Similar to the case of TYMV, RNA isolated from the swollen form of CCMV has a low sedimentation constant (7 S) and is homogeneous to a rather high degree. In both cases the molecular weight of the RNA fragmented in virions correspond closely enough with the size of the RNA fragment due to the periodicity of its tertiary structure (see above). As the TYMV RNA molecular weight is 2×10^6 and there are 32 local condensates *in situ*, the weight of the RNA fragment in such condensate should be $\frac{1}{32}$ of $2 \times 10^6 = 62 \times 10^3$. For the CCMV the expected value of RNA fragmentation *in situ* is somewhat less ($\frac{1}{32}$ of $1.09 \times 10^6 = 34 \times 10^3$). The model experiments on the degradation of RNA preparations using low levels of RNase or short alkali exposures have demonstrated a continuous distribution of molecular weight. These results suggest that the intraviral RNAs break in certain points which are distributed more or less periodically along the polynucleotide chain. Depending on the nature of the degrading agent (e.g., RNase, alkali, shearing forces), there may be various sites of rupture (within a condensate or between RNA condensates). For instance, for the alkaline degradation of TYMV the most probable sites of hydrolysis are between condensates since the RNA in such structures is in close contact with protein and is less accessible to alkali because of the steric factors and a probable hydrophobic environment (see below). Therefore, with short exposures, hydrolysis of the intraviral RNA by alkali is most probable at the sites where RNA is in the free state and is not closely associated with protein. This explanation is confirmed also by the fact that with increasing time of incubation with alkali there appear fragments of various (and, naturally, smaller) size. This is clearly due to hydrolysis of the phosphodiester bonds at the less-accessible sites blocked apparently by protein.

As we have already stressed, the above mechanism is not the only one. For instance, it may also be suggested that the interaction with protein makes some phosphodiester bonds especially sensitive to alkaline degradation (Bosch *et al.*, 1967). Also possible is the assumption put forward by Bancroft *et al.* (1968*a*) that under the conditions of a strong bond between RNA and capsomers, variation of their mutual arrangement and distances between them due to conformation alterations of the capsid may lead to ruptures of the phosphodiester bonds as a result of developing shearing forces (Kaper, 1968). Interpretation of the experimental results is somewhat hindered by the presence in the virus preparations of trace amounts of RNase, but these results cannot be fully ascribed to the presence of the degrading enzymes (cf. discussion of this problem in the earlier reviews) (Kaper, 1968; Tikchonenko, 1969).

Thus, the geometry of the RNA packing in the isometric viruses studied indicates the possibility of close contacts between the protein and certain sites of the RNA molecule *in situ*. In the RNA-containing phages the RNP is estimated to include not more than 20% of the total RNA of the virion. In the plant viruses the amount of RNA interacting with protein has not been determined. The contribution of the RNA–protein interaction to the general physical stability of the virion does not depend on the characteristics of packing of the intraviral RNA, which repeats the icosahedral symmetry of the capsid.

1.3.2. The Secondary Structure of RNA

The data discussed in the preceeding section indicate that the "amount" of the RNA–protein interactions in the viruses of this type is limited by the very principle of construction of the isometric virion, which consists of an outer protein capsid and the inner region occupied by RNA. No matter how close or strong the interactions may be between RNA and protein in the boundary zone, the zone includes only a small part of the virion RNA. In addition, the presence of the local condensates does not at all testify to the fact that all RNA in a certain condensate interacts with the protein. This fact, in conjunction with the high percentage of nucleic acid in the isometric virions, leads to a considerable part of the polyribonucleotide chain interacting "on itself."

There are direct X-ray wide-angle scattering data confirming this conclusion. The X-ray diffraction patterns of crystalline preparations of TYMV and tomato bushy stunt virus reveal reflections at 0.5 and 1.2 nm due to double-stranded parts of the intraviral RNA (Klug and Finch, 1960; Klug *et al.*, 1961). These results indicate qualitatively that a certain part of the RNA in these virions may have the same secondary structure as in solution. Up till now the unexplained result of these studies is the considerable difference in the intensity of these reflections for the two examined viruses. It is possible that despite the almost indistinguishable structure of the virions as a whole, the percentage of helical RNA structure *in situ* differs in them. However, this suggestion needs a thorough experimental substantiation since other viruses of this group do not reveal great differences in the content of ordered secondary structure of their RNA *in situ*.

Analysis of the UV absorption spectra of the intraviral RNA and its hypochromism reveals a high percentage of spiralization which does not differ in some cases from the content of the ordered secondary structure in the free RNA. Naturally, we presume here the optimum conditions (*p*H, ionic strength, and temperature) providing for the maximum stabilization of helical structure *in vitro*. Without going into the methodological aspects of the problem, which have been discussed in a previous review (Tikchonenko, 1969), we shall limit ourselves to the data presented in Table 8, as a typical example. Table 8 illustrates the calculation procedure for determining the value of the hypochromic effect, which provides the basis for evaluating the percentage of spiralization (Doty *et al*., 1959).

This calculation clearly shows that the contents of formal secondary structure in the free and intraviral RNA are practically the same. There is some uncertainty in this conclusion due to the insufficiently accurate determination of the protein contribution to the total absorption of the virion. As may be seen from Table 8, Doty and co-workers have determined it for a monomeric protein, whereas absorption by the aggregated form of protein (capsid) may be different. A similar calculation procedure has yielded similar results for some other viruses of this group: TYMV (Haselkorn, 1962) and pea enation mosaic virus (Shepherd *et al*., 1968). The same percentage of spiralization of RNA

TABLE 8

Determination of the Percentage of the Helical Structure in RNA of Wild Cucumber Mosaic Virus *in Vitro* and *in Situ*

Measurement	Virus	Free RNA
1. Initial absorption (260 nm, *p*H 7)	1.000	1.000
2. Correction for light scattering[a]	0.125	—
3. True absorption of virus (260 nm, *p*H 7)	0.875	—
4. Correction for protein contribution[b]	0.104	
5. True RNA absorption (260 nm, *p*H 7)	0.771	1.000
6. Absorption after alkaline hydrolysis[c]	1.241	1.457
7. Correction for protein contribution[b]	0.104	—
8. Hypochromism	0.68	0.69
9. Percentage of spiralization[d]	73%	70%

[a] Extrapolation procedure (Bonhoeffer and Schachman, 1960).
[b] Found for the preparation of the isolated protein.
[c] 0.3 N KOH, 24 hours, 37°C.
[d] The 0.6 value of absorption of the mixture of nucleotides corresponds to 100% spiralization at room temperature (Doty *et al*., 1959).

has also been found for the bacteriophages of the MS2 type from the UV absorption spectra and $\epsilon(P)$ values (Isenberg *et al.*, 1971).

Some authors do not determine the exact values of $\epsilon(P)$ for the intraviral RNA and use a somewhat different method for determining the degree of hypochromism of the intraviral nucleic acid; it consists of disintegrating virions and releasing the nucleic acid into solution. By varying the *p*H, ionic strength, and temperature we may influence the conditions of virion disintegration and the subsequent spiralization of the released RNA. Generally, if RNA *in situ* has a different percentage of ordered structure, the percentage of spiralization in solution is changed, affecting the value of A_{260}. As with mild methods of virion disintegration, the capsid is preserved and the problem of determination of the protein contribution is simplified though not solved in full (cf. the above problem of the mixed tryptophan–base stacking). The thermal disintegration of the virions of tomato bushy stunt virus (Bonhoeffer and Schachman, 1960) and of foot-and-mouth disease virus (Bachrach, 1964; Matheka *et al.*, 1966) in a high-ionic-strength medium is accompanied by a decrease of absorption, in contrast to its increase in a low-ionic-strength medium. Comparing these results with the behavior of the free RNA of these viruses under similar conditions, we may calculate accurately enough the percentage of spiralization. The content of the ordered secondary structure in the free and intraviral RNA of foot-and-mouth disease virus has been found to be 64% and 43%, respectively. Similar values have been obtained for the plant viruses.

Thus, the experimental data indicate that even if the interaction between RNA and protein is accompanied by variation of the percentage of spiralization in the particles of the isometric viruses, these variations are comparatively small.

We arrive at essentially the same conclusion proceeding from the results of studies of the CD spectra of the isometric viruses. The CD spectra of the intact phages f2 (Henkens and Middlebrook, 1973) and μ2 (Isenberg *et al.*, 1971) do not differ appreciably from the spectra of a mixture of the empty protein capsids and RNA. Similarly, disintegration of virions and release of RNA by short alkaline exposure do not alter the circular dichroism of the preparation. On the CD spectra of the intact and disintegrated virus three principal dichroic bands are identified, 266, 216, and 192 nm. The first band is due to RNA and the other two are due to both RNA and protein. Thus, the complete additivity of the CD spectra typical of the isolated RNA and protein indicates that the secondary structure of RNA and protein in virions is not changed as compared to the isolated components. Calculations based on the CD spectra have yielded the value of 63 $\pm$ 5% for the content of formal secondary structure in intraviral RNA of the phages f2 and μ2 *in situ* and *in vitro* (Henkens and Middlebrook, 1973; Isenberg *et al.*, 1971). It has been repeatedly suggested that in the isometric viruses the interaction between RNA and protein may take place at the expense of the nonspiralized parts so that this process cannot be registered using the UV spectrophotometry. The identity of the CD spectra for the intact and disintegrated virus preparations (the amplitude of the CD band at 260 nm) suggests that no significant disturbance of the intra-stacking interactions occurs in these single-stranded parts.

Along with variation of the percentage of ordered secondary structure, RNA *in situ* may be characterized by a greater or lesser stabilization at the expense of both the interaction with protein and the purely steric factors connected with the conditions of RNA immobilization when it is closely packed. The reported data in this field suggest that the viruses studied may be divided into two classes differing in the stabilization of their virion RNA. Usually, in studies of this kind only the thermal stability is analyzed by comparing the melting curves for the intraviral and free RNA. This technique allows, as a rule, the comparison of only the initial portions of the curve, since at "the most interesting moment" the virus is destroyed by heating and its RNA is released into solution. It has been found for many plant viruses and for the foot-and-mouth disease virus that when a suspension of viral particles is heated the absorption does not change markedly until the viral particles begin to disintegrate. Therefore, comparing the melting temperature of the free and intraviral RNAs we may come to a conclusion about the considerable thermal stabilization of its helical parts in the latter case (Bachrach, 1964; Matheka *et al.*, 1966). In this respect the isometric viruses also resemble ribosomes (Shatsky *et al.*, 1971).

The RNA-containing phages exhibit different behavior. The initial parts of the melting curve of the RNA of the phage μ2 in the free state and in virions prove to be identical and vary similarly with the variation of the ionic strength of the medium (Isenberg *et al.*, 1971). The causes for the lack of thermal stabilization of the RNA in the phage virions are not yet known.

The thermal stabilization of the RNA of viruses may be altered by mutational changes and such poorly characterized processes as the adaptation to a host. Thus, Matheka *et al.* (1966) obtained two clones of the foot-and-mouth disease virus in the process of its adaptation to calf kidney cells. These virus clones differed in electrophoresis and in thermal stability. The authors believe that the strain with the higher electrophoretic mobility is characterized by weakened interaction between RNA and protein.

As the secondary structure of RNA in the isometric viruses has been shown by optical studies to be rather similar to the conformation of free-RNA preparations in solution the reactivities of their functional groups should also be similar. Unfortunately, the methods of chemical modification have rarely been used for studying isometric viruses.

The rate of deamination of the bases in the free RNA of southern bean mosaic virus and in the virions have proved to be almost the same (Table 9). The similar rates of deamination indicate that base amino groups are involved in complementary interactions to the same extent, i.e., that the percentage of spiralization of this RNA *in situ* and *in vitro* is the same (Sehgal and Krause, 1968; Sehgal and Soong, 1972; Sehgal, 1973). At the same time, the inactivation rate upon deamination of the intact virus is twice that for the free RNA. A similar difference between the inactivation rates has been found for poliovirus and its free RNA (Boeye, 1959), although in this case there are no data on the kinetics of base deamination. The higher inactivation rates of the intact virus as compared to the infectious RNA may be associated with reactions of the proteins and possibly with cross-linking of RNA and protein. There are no direct data on cross-linking in the particles of southern bean mosaic virus, but a highly suggestive observation has been made that the protein content in the RNA preparations isolated from the deaminated virus is 5–10 times that of the control (Sehgal, 1973).

Hydroxylamine treatment results in similar inactivation of the poliovirus and its free RNA (Gendon, 1966).

Finally, we have mentioned already the report of Turchinsky *et al.* (1974), who studied the cross-linking of a small number of protein subunits to the RNA of the MS2 phage when treated with bisulfite. The

TABLE 9

Deamination of the Free and Intraviral RNA of Southern Bean Mosaic Virus

	Percentage of deamination[a]			
	Intact virus		Free RNA	
Bases	48 hr	96 hr	48 hr	96 hr
Adenine	24.1	51.7	25.0	50
Guanine	19.2	45.0	19.8	43
Cytosine	18.2	37.3	17.2	36.7

[a] The percentage of deamination is determined from the decrease of the respective bases.

limited amount of protein cross-linked to RNA under these conditions may be attributed to the limited contacts between the bases and the nucleophilic amino acids of the capsid.

The results obtained by Werbin *et al.* (1968) and Yamada *et al.* (1973) for UV irradiation of f2 and MS2 phages, respectively, may be explained in a similar way. In spite of unchanged adsorption, penetration of the RNA into the cell was heavily suppressed upon UV irradiation. Heavy doses of UV radiation decreased the transfer of RNA from the adsorbed phage to the cell to 10% of that of the untreated phage. However, no direct evidence in favor of RNA–protein cross-linking has been reported.

Interesting information pertinent to the conformation of the RNA in isometric viruses may be obtained from UV-irradiation experiments, although the results obtained by various authors often differ. For example, the UV sensitivity displayed by the intact f2 or MS2 phages and their free RNAs is the same according to Furuse and Watanabe (1971). Similar results have been obtained by Sehgal (1973) with southern bean mosaic virus and Kassanis and Kleczkowski (1965) with tobacco necrosis virus. This viewpoint has been confirmed to some extent by the deuterium isotope effect values, that is, the ratios of the doses required to inactivate the viruses to the same survival levels in D_2O and H_2O. The isotope effect values for MS2 and Qβ phages proved to be 1.1 and 1.2. The same values have been obtained for the free RNAs of these phages (Furuse and Watanabe, 1971). These results indicate similar hydration of RNA *in vitro* and *in situ* (see above).

In contrast to this, Werbin *et al.* (1967) and Yamada *et al.* (1973) have found a distinct difference in the UV sensitivity of free and encapsidated RNA of R17, MS2, and f2 phages. Besides that, pyrimidine dimers are completely absent in the UV-irradiated R17 and MS2 phages, although they are easily formed in the free RNA preparations (Remsen *et al.*, 1970, 1971; Yamada *et al.*, 1973; Mattern *et al.*, 1972). This fact agrees with the well-established inability of the UV-irradiated RNA-containing phages to be photoreactivated (Rauth, 1965; Zinder, 1965). At the same time, free RNA of R17 irradiated *in vitro* can be reactivated in bacterial protoplasts (Werbin *et al.*, 1967; Furuse and Watanabe, 1971). One may easily imagine that the absence of the pyrimidine dimers in the phage, on one hand, and their formation in free RNA, on the other hand, result in different inactivation. Finally, in the UV-irradiated phage one may expect formation of cross-links between RNA and protein (see above).

In comparing these contradictory results obtained by various authors using different virus models, one must keep in mind two things.

First, the equal inactivation of free and encapsidated RNAs of plant viruses may be explained to a certain extent by the low sensitivity of the bioassay methods of these viruses. Second, it is easy to overestimate the usefulness of the data yielded by the deuterium isotope effect experiments. The individual and poorly understood macromolecular properties of different RNA species may influence to a great extent the isotope effect values. For example, for free TMV RNA this value is 2.0 (Tao *et al.*, 1966), in contrast to the value of 1.2 found for the RNA of Qβ phage (Furuse and Watanabe, 1971).

In the section dealing with the rodlike viruses we discussed the possible relationship between the secondary structure of RNA and formation of pyrimidine dimers following UV irradiation. Therefore, the absence of these dimers in the UV-irradiated isometric phages seems to contradict the above viewpoint. Trying to eliminate this discrepancy, Remsen *et al.* (1970) have put forward the suggestion that sufficient mobility of the residues in a polynucleotide chain, allowing the orientation of neighboring pyrimidine rings into a geometry favorable for cyclodimerization, may be an important prerequisite for the reaction to proceed efficiently. RNA in isometric phages may be held in a rigid conformation in intimate contact with the phage protein [see Klug *et al.* (1966*a*)]. Alternatively, suppression of cyclodimerization may result from a quenching effect of the phage protein on the excited state of the RNA responsible for dimerization but not for photohydration. At the same time, the *total* absence of pyrimidine dimers is in poor agreement with the *partial effects* which must be observed in the latter case.

Certain differences in the reactivities of the bases of intraviral RNA and RNA *in vitro* may be revealed only by means of mild-acting agents highly sensitive to base stacking and complementary interactions. Budowsky *et al.* (1972) have demonstrated by their careful analysis that inactivation and mutagenesis caused by O-methylhydroxylamine (OMHA) have different kinetics for the intact phage MS2 and its free RNA. In contrast to the intact phage, when 0.2 M OMHA acts on the infectious RNA there is a certain lag phase before the beginning of mutant accumulation, the plateau on the mutation curve for RNA appears somewhat earlier than for the phage, and the height of this plateau is half that for the intact virus. These differences have not been observed with higher OMHA concentrations. The differences found between the phage and its isolated RNA in the course of inactivation and mutagenesis have not been interpreted as yet but, undoubtedly, they indicate some effect of the RNA conformation *in situ* on the course of its reaction with a chemical mutagenic agent.

1.3.3. RNA–Protein Interaction in Virions

The interactions between protein and RNA in the isometric virions are described by a comparatively large amount of experimental data, but, unfortunately, the data are of predominantly indirect character. When the MS2 phage is modified by bisulfite (Turchinsky *et al.*, 1974) cross-linking of protein and RNA indicates the involvement of the cytosine-residue C^4 atom in some interaction with protein.

The data on the RNP interactions in these viruses are mainly concerned with the salt bonds. The results of the experiments on reconstitution of these viruses demonstrate the existence of such bonds between RNA phosphate groups and basic amino acids in viral protein. As discussed above, the isometric viruses are characterized by different contributions of protein–protein and RNA–protein interactions to the general physical stability of the virion. It is just the fact that the protein subunits in the viruses of the first group are capable of strong and specific interactions that allows their capsids to be formed *in vitro* by means of self-assembly of the protein subunits not involving RNA. The viruses of the second group are not capable of capsid self-assembly in the absence of RNA, at least under experimental conditions which provide for assembly of the protein shell in viruses of the first group. However, it should be noted that the capsid self-assembly from the protein subunits in the absence of RNA does not always yield unambiguous results in the plant viruses belonging to the first group (Bancroft *et al.*, 1968*a,b*; Hiebert *et al.*, 1968). Aggregation of the protein subunits may give rise to various capsids differing in size and form, with only a small percentage of the typical capsids of mature virions. In the presence of RNA reconstitution yields normal capsids. Protein–protein interactions also affect capsid formation for the RNA-containing phages (Hohn, 1969; Hohn and Hohn, 1970). The refined data (Matthews and Cole, 1972; Schubert and Frank, 1971) show that normal phage capsids may be formed in the absence of RNA provided that the ionic strength of the medium is high enough and the temperature is lower than 25°C. The physical stability of such capsids does not differ from that of the protein shells formed by virion reconstitution, i.e., in the presence of RNA. Analysis of the phage capsid degradation indicates that the presence of RNA little affects the stability of the protein shell. Reconstitution of virions requires polyanions rather than RNA; therefore, the isometric viruses have been successfully reconstituted not only in the RNA–protein mixture but also in mixtures of protein and polyvinyl sulfate (but not heparin). Of certain importance for reconstitution is the cooperativity of the salt bonds determined by the polymeric state of the RNA or polyanion. In particular, the capsid assembly proceeds only in the presence of polynucleotides containing no less than 20 residues. The composition of the bases and replacement of phosphate by sulfate do not decisively affect the assembly results, but they influence the rate of the process (Hohn and Hohn, 1970). However, under the conditions of strong and specific protein–protein interactions polyanions may be replaced by high salt concentrations in solution.

Thus, the experiments on the capsid self-assembly and on virion reconstitution and analysis of the stability of the protein shell suggest that the principal structural function of RNA in the isometric virions with predominant RNA–protein interactions is the neutralization by means of salt bonds of the positively charged protein groups localized at the capsid's inner surface (Hohn and Hohn, 1970; Matthews and Cole, 1972).

Some authors (Heisenberg, 1966; Hohn, 1969; Hohn and Hohn, 1970) proceed from the experimental results on *in vitro* phage reconstitution, attempting to determine quantitatively the RNA involvement in the salt interactions. These calculations proceed from the fact that in the absence of the maturation factor (protein A) in the medium or in the infected cell, noninfectious particles are formed whose RNA is not completely covered by capsid protein. The self-assembly involving the heterologous RNA gives rise to defective virions whose buoyant density is higher than that of the normal phage. When such particles are treated with RNase, the extracapsidal excess of RNA is eliminated and the buoyant density of the virions goes down to 1.38 g/ml. This value corresponds to the buoyant density of particles containing two-thirds of the normal amount of RNA. Noticeable here is the close similarity between the amount of basic amino acids in the capsids of the f2 and R17 phages and the amount of RNA left in the defective virions after elimination of that part of the RNA which was accessible to RNase. Two-thirds of the normal amount of RNA corresponds to 2000 nucleotides, whereas the protein of the phages R17 and f2 contains 1980 and 1800 arginine plus lysine residues, respectively. Hohn (Hohn, 1969; Hohn and Hohn, 1970) has suggested that most of these basic amino acids are involved in the salt bonds with RNA, thus protecting the main part of this RNA from degradation caused by nucleases in the particles lacking the A protein. Similar defective virions containing two-thirds of the normal RNA content are also isolated from cells infected with some amber mutants of the f2 and fr phages (Heisenberg, 1966).

Thus, the salt bonds required for successful capsid assembly are retained in the virion after its formation. Yet, a comparison of these results with X-ray diffraction data on the RNA–protein "contacts" indicates a disagreement. If we proceed from the assumption that no more than one-fifth of the RNA is in contact with protein (see above), i.e., 600 nucleotides, the figure of 2000 nucleotides evidently seems to be an overestimation.

The interaction between RNA and protein in the poliovirus nucleocapsid is of considerable interest in this conceptual framework since there are good reasons to assume that the first to be assembled *in vivo* is the poliovirus protein capsid, with later addition of RNA (Jacobsen and Baltimore, 1968). Clearly, the salt interactions between RNA and protein are not a precondition for the successful capsid self-assembly for this isometric virus and, possibly, not for any enteroviruses.

However, the number of salt interactions between RNA and protein and their contribution to the general physical stability of virions differ significantly in various virus groups. Kaper, Hirth, and Bancroft have obtained the most complete data on TYMV and CMV, which represent the virus groups with strong and weak protein–protein interactions, respectively (see above).

Degradation of TYMV by means of alkali (Bosch *et al.*, 1967; Kaper, 1968; Kaper and Halperin, 1965), urea, and formamide (Jonard and Hirth, 1966; Jonard *et al.*, 1967; Bouley and Hirth, 1968) depends to a great extent on the ionic strength of solution. We shall not discuss here the direct stabilizing effect of high salt concentrations on the protein capsids [see Kaper (1968)], instead we will limit ourselves to the results describing the RNA–protein interaction. Prolonged incubation of TYMV in alkali in the presence of 0.5–1 M KCl leads to formation of empty capsids and fragmented RNA. Undamaged capsids and degraded RNA may also be obtained by short-time heating of the virus suspension in 1 M KCl at 65°C and *p*H 6.0.

Yet, if the ionic strength is low (0.05 M KCl), the alkaline incubation leads mainly to the decomposition of the capsid into low-molecular-weight fragments, although a considerable part of the capsid is retained in the process. In contrast to the intact virions, the alkaline resistances of the empty protein capsids in high- and low-ionic-strength media have proved to be fairly similar, if not the same, and uniformly high (Kaper, 1968). Similarly, incubation of TYMV with 8 M urea or 12 M formamide in the high-ionic-strength media is accompanied by the prevailing formation of empty protein capsids, whereas in the low-ionic-strength media the capsids are decomposed to lower-molecular-weight material consisting of aggregates of peptide chains of varying length (there is no RNA degradation in the process) (Cantor *et al.*, 1966; Hill and Shepherd, 1972; Damirdagh and Shepherd, 1970).

Thus, the presence of RNA in the TYMV nucleocapsid is associated with the lower capsid stability when it is degraded by alkali or urea; this is especially noticeable in low-ionic-strength media. These results are sufficiently unambiguous to suggest the existence of both H pairing and salt interactions between RNA and protein involving, in the latter case, phosphate groups, on one hand, and basic amino acids, on the other. At alkaline *p*H values both salt and H bonds should be ruptured due to the loss of charge by the protonated amino acids, whereas urea and formamide should rupture only the H bond. Both these agents totally or partially disrupt RNA–protein interaction, leading to the release of RNA into the medium. As this process does not depend on the ionic strength of the medium and on the preservation of the protein capsid, clearly, the release of RNA into the medium is determined by the nucleic acid itself. Evidently, the determining factor is the difference in the volumes occupied by the RNA molecule in the compact configuration *in situ* under the conditions of interaction with protein and in the free state. When the interaction with the protein is disrupted, the tertiary structure of the molecule becomes more loose due to the electrostatic repulsion of the phosphate groups, with highest ionization at *p*H 10–11, leading to a sharp increase in the molecular volume and to the rupture of the capsid. In the low-ionic-strength media, where the shielding effect of cations with respect to phosphates is minimized, the molecular-volume difference between the two RNA conformations should be especially high and the RNA release into the medium occurs in an explosive manner.

Proceeding from the studies of TYMV degradation under various *p*H and ionic-strength conditions, Kaper (Kaper, 1969, 1971, 1973; Kaper and Geelon, 1971) has shown experimentally that RNA–protein bonds rupture in the *p*H range 6–7. Proceeding from this result and from the known *pK* values for various functional groups of RNA and protein, it may be suggested that the RNA–protein interaction in this virus is represented by either ionic (salt) bonds between protonated hystidine residues and negatively charged phosphate groups or by H bonds between carboxyl and RNA amino groups. In TYMV this role may be performed by the cytosine amino group since the content of cytosine in the RNA of this virus amounts to 38 mol.%.

The successful recent *in vitro* assembly of TYMV-like particles from RNA–protein mixtures apparently allows a choice to be made between two possible types of RNA–protein interactions (Jonard *et al.*, 1972) in the virions of this group. This reaction is carried out with decreasing salt concentrations and in the presence of $MgCl_2$ or spermidine, with the necessary condition that the *p*H value does not exceed 4.5. When the *p*H is increased up to the neutral value the reassociated RNP particles decompose, even though intact TYMV is stable under these conditions. Nevertheless, these assembly conditions suggest that the principal type of RNA–protein interactions in an isometric virus of the TYMV type is H bonding involving protein carboxyl groups. Even though these bonds are not the main source of the virion's physical stability, the interaction with RNA is necessary for formation of viruslike particles. It has been suggested that despite the strong protein–protein interactions and the similarity between the structures of the empty capsids and the capsids in virions, the protein conformations in them differ somewhat under the influence of RNA (Kurtz-Fritsch and Hirth, 1972).

In the CMV group, including brome mosaic virus, broad bean mottle virus, and others, the main type of RNA–protein interaction is the ionic bond between RNA and protein, and it is at the same time the main factor in the physical stability of these virions due to weak protein–protein interactions (Kaper, 1973). This conclusion is, primarily, substantiated by the easy dissociation of these viruses into RNA and protein subunits in 1 M salt solutions at room temperature. However, if the ambient salt concentration in the mixture of protein and RNA (or DNA and even polyvinyl sulfate) is decreased, they are easily reassociated (Hiebert *et al.*, 1968; Bancroft *et al.*, 1969). Dissociation of CMV in concentrated salt solutions is independent of *p*H over a wide range, indicating the highly acidic and highly basic nature of the dissociating groups giving rise to the ion pairs. This is the case for the RNA phosphate groups and the basic protein amino acids.

Beside the fact that part of the RNA phosphates should be involved in salt interactions with protein, it may be suggested that some RNA phosphate groups *in situ* may be in a nondissociated state (Kaper, 1968). It is known that in TYMV the total amount of basic groups of the virus protein and polyamines present in the virion does not suffice for neutralization of all 6500 phosphate groups of the nucleic acid. Even though neutralization of RNA phosphates *in situ* is provided by Mg^{2+} ions, Kaper suggests that part of the phosphates in a nonpolar environment, i.e., in a medium with a low dielectric constant, may be nondissociated. This is why TYMV decomposition in the process of incubation with *p*-chloromercuribenzoate in low-ionic-strength media leads to strong acidification (Kaper, 1968). The capsid decomposition is assumed to lead to the disappearance of the nonpolar environment of the phosphate groups so that they become capable of normal dissociation, providing for the rapid release of hydrogen ions.

Similarly, the potentiometric titration of empty protein capsids and intact TYMV has shown that the amount of bound alkali in the second case significantly exceeds the amount of NaOH bound by the capsids (Kaper, 1968). The author's calculations have demonstrated that this difference cannot be ascribed only to titration of

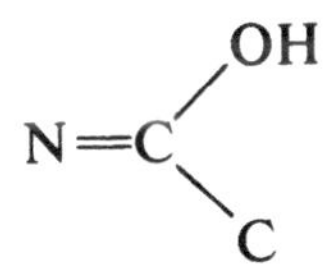

groups of uridylic and guanylic acids with *pK* values of 9.5. Kaper explains this result by other prototropic nucleic acid groups which are titrated in regions of more acidic *p*H under normal conditions. However, these groups become inaccessible for titration in the virion due to the close interaction with protein, and they lose their protons only after certain conformational alterations induced in the protein during alkaline dissociation. Under these conditions the apparent *pK* values of these groups are shifted to the highly alkaline region. This approach seems to be of great interest and may contribute significantly to our knowledge of the RNA structure *in situ* and its environment. However, parts of such groups with pseudo-alkaline *pK* values might belong not only to RNA but to protein as well if the subunit conformation differs in the free state from that in virus.

1.4. Isometric DNA-Containing Viruses

In comparison to the RNA-containing viruses, this problem is yet to be studied for the viruses containing single-stranded DNA. The few experimental results concerning DNA structure *in situ* have been obtained when studying other problems. According to these fragmentary data, the structure of the DNA in the isometric viruses differs little from the conformation of this nucleic acid in solution. The DNA in the phage ϕX174 (Sinsheimer, 1959) and in the minute virus of mice (MMV) (Crawford, 1966) reacts at 37°C with 1% HCHO, giving the same final 25–26% of hyperchromism as the free DNA preparations.

No special studies of the chemical modification of the DNA of such viruses *in situ* have been conducted, though ϕX174 and related phages have been extensively used as models when studying chemical mutagenesis. However, no significant differences have been found in the experiments with HNO_2 between the inactivation and mutagenesis curves for the intact viruses and the infectious DNA preparations (Belych and

Krivissky, 1966); nor have significant differences been reported between the inactivation kinetics of intact ϕX174 phage and its infectious DNA upon incubation with OHMA (Krivissky *et al.*, 1973).

There is no agreement between the results of various studies of the effect of UV irradiation on the phage and its isolated DNA. Belych *et al.* (1968) reported that 253.7-nm-wavelength UV light inactivated intact ϕX174 more effectively than its infectious DNA. Yet, the mutagenic effect of UV irradiation is greater in the intact phage than in the DNA preparations. However, as reported by Dityatkin *et al.* (1967), the sensitivity of the closely related phage ϕL-7 and its DNA to 253.7 nm UV irradiation is the same. These authors have found some differences in the UV sensitivities of this phage and its DNA at the 265 nm wavelength which seem to be due to the damaging action of radiation on the protein component. No explanation has been found for the different UV sensitivities of the phages ϕX174 and ϕL-7.

No detailed CD and ORD studies have been performed on viruses with single-stranded DNA. CD and ORD spectra have been obtained only for the intact phage ϕX174 and its DNA preparations (Minchenkova *et al.*, 1969; Maestre and Tinoco, 1967), but the lack of the respective data for the empty protein capsids hampers interpretation of the results. According to the ORD studies of the complex of the intraphage DNA with Ag^+, the degree of DNA spiralization for the ϕX174 phage *in vitro* and *in situ* is the same (Minchenkova *et al.*, 1969).

Thus, the few and fragmentary data on the isometric viruses with single-stranded DNA fit into the conceptual framework proposed for the RNA-containing viruses; the interaction between DNA and protein in virions of this type is limited and does not lead to any significant alteration of DNA conformation *in situ*.

2. VIRUSES WITH DOUBLE-STRANDED DNA

This section is limited to the bacteriophages with isometric heads containing double-stranded DNA, as there are no data at all on other viruses with this type of nucleic acid and no data on the structure of double-stranded RNA *in situ*.

2.1. The Tertiary Structure of DNA

In all the known bacteriophages the length of the DNA molecule is 300–500 times greater than the head diameter [see references in Tikchonenko (1969)]. For this awkward, long object to be accommodated in a limited spherical volume, it must be made more compact. There are two theoretically possible ways of solving this problem: to pack the DNA in the phage head in some order or to push the DNA into the head in a completely chaotic way, much as a lazy housewife might stuff her linen closet, her only objective being that no linen be left outside. The second type of packing, widespread as it may be in householding, has no place in the domain of viruses since Mother Nature is extremely tidy and keeps strict order in everything, including DNA packing in the bacteriophage heads. In addition to "keeping up with the Joneses," nature has another reason for turning to the ordered ways of packing, namely, that chaotic packing is essentially inconsistent with the crowded housing conditions for DNA in the phage particles, on one hand, and with the high speed of DNA injection into the infected cell, on the other. The first of these two factors has been discussed in detail in an earlier review (Tikchonenko, 1969). Generally, it means that viruses on the whole and bacteriophages in particular, for reasons of economy elevated to the rank of biologic expediency, have to be content with extremely limited space in the capsid accommodating the DNA. The ratio between the virion's inner cavity volume and the DNA molecular volume evaluated by various methods is close to unity, indicating extremely compact and economical packing. As an example, let us consider the best-studied phage, T2, which has the form of an icosahedron with an elongated middle part (Moody, 1965; Boy de la Tour and Kellenberger, 1965). The volume of such a body may be found from the following formula:

$$V = 0.718b^2L - 0.322b^3$$

where L is the head's length, or its maximum diameter, and b is its thickness or the minimum diameter (Klimenko *et al.*, 1967). The diameter of the T2 phage head determined in negatively stained preparations is 119 × 90 nm according to the most generous estimates and the minimum thickness of the capsid membrane is 3.5 nm (Bradley, 1965). The inner diameter of the capsid determining the inner cavity volume (V_i) is, thus, 115.5 × 86.5 nm.

Calculations of the volume available for the DNA (V_D) usually proceed from the so-called packing reflection of 2.4 nm, which is regularly observed on the X-ray diffraction patterns for the aligned DNA preparations and is also found on the X-ray diffraction patterns for the intact viruses (Dembo *et al.*, 1965; North and Rich, 1961). This reflection is interpreted as the minimum distance between the centers of the parallel rows or segments of the double-stranded DNA. The validity of such treatment is confirmed by the results of electron microscopic studies of the DNA condensates released from the phage heads upon capsid rupture (Klimenko *et al.*, 1967; Richards *et al.*, 1973). According to these micrographs, the distance between the centers of the parallel DNA strands is 2.5–3.0 nm. Therefore, the volume occupied by DNA with such packing should be calculated from the following formula:

$$V = [2/(3)^{1/2}]D^2L$$

where D (diameter) is 2.4 nm and L (length) is 62×10^3 nm for the T2 phage (Thomas, 1963). The results of these calculations are summarized in Table 10.

Admittedly, Table 10, and particularly its bottom row, must seem to be strange for an unprepared reader. The ratio $V_i/V_D = 0.87$, which looks quite respectable in a scientific text, should make a stunning impression when translated into conventional language since it describes the absurd situation when one thing located inside another thing cannot conceivably be accommodated there. In addition, the ratio might be even lower as the above calculation is based on the maximum possible head size and the minimum membrane thickness [for references see Tikchonenko (1969)]. This logically impossible situation may be accounted for in a number of ways. First, we may assume that the secondary structure of the DNA *in situ* differs to some extent from its structure in solution; therefore, calculation of V_D based on the similarity to the native DNA of the Watson–Crick double-helix form is arbitrary. These were the author's foundations for postulating the existence *in situ* of a partial disturbance of the ordered secondary structure of the double helix (Tikchonenko *et al.*, 1966). Discussing the possibility of the C form of DNA in phages and nucleoproteins suggested by Maestre (Tunis-Schneider and Maestre, 1970; Dorman and Maestre, 1973; Holzwarth *et al.*, 1974). Nelson and Johnson (1970) have suggested as a rational justification for minimum DNA packing that C-form DNA has the least volume in comparison to the A and B forms. Indeed, in orthorhombic C form fibers each nucleotide occupies only 10.7 nm³ *vs.* 11.8 nm³ in the A form, B form, or hexagonal C form of DNA (Marvin

TABLE 10

DNA Packing in the T2 Phage Particles

Phage head volume	$480\ nm^3 \times 10^{-5}$
Inner cavity (V_i) volume	360
DNA molecule (V_D) volume	413
Ratio V_i/V_D	0.87

et al., 1961; Fuller *et al.*, 1965; Langridge *et al.*, 1960). Second, as illustrated by the RNA-containing phages, the regions occupied by DNA and protein are not distinctly separated by a border. Clearly, there exists a boundary region where mutual penetration of DNA and protein is allowed. This "dual nationality" possibility must evidently increase the value of V_i.

Another factor which also requires a regular packing of DNA in the phage head concerns injection of viral DNA into the infected bacterial cell. The needle channel through which DNA is transported from the phage head into the cell has a diameter of 2.0–2.5 nm, i.e., practically equaling the molecular diameter of double-stranded DNA. Therefore, irrespective of the mode of packing of the phage DNA *in situ*, it may get into the infected cell only in the form of an elementary double strand, i.e., its tertiary structure should undergo the transformation which in the parlance of the scientists is called "unwinding." Despite the fact that various authors have different viewpoints concerning the DNA injection mechanism (Zarybnicky, 1969; Zaretsky *et al.*, 1971; Perera and Tikchonenko, 1969; Mekshenkov and Guseynov, 1971; Manykin, 1972), they all agree that this process occurs very rapidly, and, in particular, it takes less than a minute for the T2 phage at room temperature. The figure quoted here (1 minute) is only the upper limit since the measurement technique used for determination of the injection rate does not allow the measurement of shorter time intervals. Therefore, packing of the intraphage DNA must permit high-speed unwinding or decondensation.

There are three methods of compact, regular packing the DNA strand into the phage head: as a pile of rectilinear segments, by circular winding (as a ball of thread or as a spool of thread) and, as usual, a compromise way combining features of the preceeding two methods. In the first case the DNA strand is repeatedly bent by 90°, forming rows of antiparallel segments whose length is determined by the head's diameter. In the second case, the individual double-stranded DNA is circularly packed without bending, forming finally a hollow sphere, a ball, or a spool. In the compromise version it is assumed that this spherical shape is formed not by an individual DNA strand but by a thread formed of bent DNA segments. The first concept was supported by a large number of authors during the 1960s (Bendet *et al.*, 1960; Bendet, 1963; North and Rich, 1961). The second model has been developed by Richards *et al.* (1973); the third has been developed by our group (Klimenko *et al.*, 1967; Tikchonenko, 1969). The final choice among the three models depends to a great extent on the results of studies of the variously oriented DNA preparations.

A considerable degree of asymmetry of many phages due to the long tail enables the scientist to obtain oriented particles by centrifugation in narrow capillaries, by pulling threads out of the phage sediments, under the action of the surface-tension forces, or by using flow methods and electric fields in birefringence experiments [for references see Tikchonenko (1969) and Greve and Blok (1973)]. Except for the two last methods, there are yet no criteria for reliably estimating the degree of particle orientation in a preparation. Nevertheless, the flow birefringence for the oriented phage T2 particles has proved to be negative, while it is positive for the empty phage shells (ghosts) (Bendet, 1963, Gellert and Davies, 1964). This has led to the conclusion that the negative birefringence is due to the DNA packing *in situ* and its partial orientation with respect to the virion axis. Later, Maestre (1968) and Greve and Blok (1973) used electric birefringence at high field strengths (saturation region), where the orientation of T2 and T4 particles was nearly 100% parallel to the applied field. Practically, the orientation function (ϕ) in their experiments was equal to 1. These later studies allowed the elimination of the vagueness of early experiments and made possible a more reliable estimation of DNA orientation *in situ*.

In early studies the authors proceeded from the parallel-segment packing model for DNA *in situ*. In this model the segments are oriented with respect to the long axis of the particle. The first semiqualitative studies were concerned with the predominantly parallel orientation of the DNA segments with respect to the particle axis, but it had been suggested in the later studies that the degree of orientation was not high and that for T2 and T4 phages it did not exceed 9% (Gellert and Davies, 1964). Later this value was increased up to 20% (Maestre, 1968). Special attention has been paid in recent studies to the presence of free DNA released from the phage particles. The precautions employed to prevent the release of DNA into solution permit the safe conclusion that the specific birefringence observed is not due to the released DNA. The latest results obtained by O'Konski, Kwan, Shepherd, and Maestre (in preparation) demonstrate negative birefringence for T2, T4 "0," and T5 phages and positive birefringence for T7 virions. The cause of these differences in the birefringence sign between different phages is not yet properly understood.

The X-ray diffraction results for the threads obtained from the fresh sediments of the "tailed" phages proved to be no less inconsistent. The early reports suggested a high degree of DNA orientation with respect to the strand axis on the basis of the X-ray diffraction results for such oriented preparations as compared to nonoriented gels (North and Rich, 1961; Kilkson and Maestre, 1962), whereas later studies led to opposite conclusions (Dembo *et al.*, 1965).

Dembo *et al.* (1965) have compared the possible degree of DNA orientation in the head of the perfect isometric DDVII phage and in the head of the T2 phage, the capsid of which has icosahedral symmetry with an elongated middle part. The degree of orientation was estimated from intensification of the reflection at 0.34 nm in the phage threads as compared to the nonoriented gels of viral particles. Even though it is difficult to make accurate estimations in such experiments, some differences were found between the phages studied. Only very slight intensification of the reflection in the direction of the particle axis was observed for the DDVII phage, whereas for the T2 the reflection intensification was greater (Dembo *et al.*, 1965).

There are two possible explanations of these results. First, intensification of reflection due to orientation in some phages may be associated with the presence of free DNA in the phage threads because it is very difficult in the X-ray experiments compared with birefringence studies to perform adequate control tests for free DNA (see below). This being so, the difference observed between the T2 and DDVII phages should be ascribed to the fact that the former phage is labile and readily loses its DNA whereas the latter phage is more stable. Second, intensification of reflections in the T-even phages may be due to the characteristics of intraphage DNA packing, not as straight, antiparallel segments but as rings.

In addition, the above-mentioned results obtained by O'Konski, Kwan, Shepherd, and Maestre (in preparation) testify to the possibility of a different arrangement of DNA in T-even and T7 phages. The same may also be true for phages of the DDVII type.

As for the first possibility, usually in these cases a study is made of the reflection behavior near 2.4 nm, depending on the relative humidity of the specimen. As this reflection is due to scattering on the parallel aggregated DNA strands or its segments, the reflection at 2.4 nm should disappear due to the dissolution of the specimen at 100% humidity. This disappearance is, indeed, observed in the oriented samples of the deproteinized DNA. However, nucleoproteins, including artificial DNA–protein complexes, are distinguished by strong bonds between aggregated DNA strands (Thomas, 1963). Manipulation of the oriented phage particles, especially the labile phage T2, may result in partial release into the medium of DNA complexed with the internal protein. The latter is known to possess basic properties and to cause intense DNA aggregation (Chaproniere-Rickenberg, 1964; Bachrach and

Friedmann, 1967). Therefore, the preservation of the equatorial scattering near 2.4 nm in the oriented phage preparations at 100% relative humidity found in the quoted studies may also be ascribed to free DNA. A combination of the flow birefringence data and X-ray scattering data indicates that there is no preferred orientation of the DNA segments with respect to the long axis in the phage particles studied. If some DNA orientation with respect to the T-even phage particle axis observed by Gellert and Davies (1964) and Maestre (1968) is not due to the presence of free DNA in the preparation of oriented molecules, the above differences in behavior between the T-even and DDVII–T7 phages may be used to solve finally the problem of the intraphage DNA packing. For the sake of discussion continuity we shall expound these considerations later in connection with the analysis of the electron microscopy results.

The presence of the ordered tertiary structure of DNA *in situ* is evidenced by the above equatorial scattering near 2.4 nm (the so-called "packing reflection") and by the results of small-angle scattering studies (Dembo *et al.*, 1965; Katz and Rich, 1966). In this case the mean size of the regular packing region was determined, using the Debye–Scherrer method, from the diameter of the reflection ring corresponding to the Bragg reflection near 2.4 nm. This diffuse maximum of the X-ray scattering curves may be interpreted either as a reflection due to an ordered packing region ("crystalline domain") or as a certain parameter of disordering in the system defined as the interaction radius (Vainstein, 1963). In the first case we mean an autonomous ordered region in the traditional sense of the word, for instance, a real thread of molecules or of segments of one molecule. In the second case the molecular interaction is treated in a rather formal way and this parameter describes not a real region in the regular packing zone but disturbances of ordered arrangement of the DNA segments in this zone as a function of distance from a certain arbitrary center. However, irrespective of the physical interpretation, the quantitative expression of this parameter is the same. Meanwhile, the electron microscopy results (see below) demonstrate the existence *in situ* of ordered aggregates of DNA segments. This finding favors the first viewpoint. Dembo *et al.* (1965) and Katz and Rich (1966) have reported similar sized crystalline domains, varying from 10 to 14 nm for a number of large and medium-sized phages. Thus, some indirect evidence leads to choosing the first viewpoint and assuming that the tertiary structure of the intraphage DNA has a certain suborganization in the form of ordered aggregates 10–14 nm in diameter consisting of the DNA segments.

It should be admitted that the ambiguity of the available experimental data makes this viewpoint open to criticism. In particular, Richards and co-workers (1973) do not recognize suborganization of the intraphage DNA packing. They believe that the tertiary DNA structure *in situ* is limited to the ordinary circular winding of an individual strand of the double-helical DNA.

The presence of suborganization in the tertiary structure of the intraphage DNA has been recognized by Kilkson and Maestre (1962), Maestre and Kilkson (1962), Rubinstein (1960), Klimenko *et al.* (1967), and Tikchonenko (1969). In the early 1960s the intraphage DNA was assumed to be organized in the form of a secondary helix with a diameter of 7.5 nm or less (Kilkson and Maestre, 1962). This assumption seems to have been disproved by the data on the tertiary structure of the condensed DNA demonstrated by electron microscopy of the phage particles disintegrated under certain conditions (Klimenko *et al.*, 1967; Richards *et al.*, 1973). The ring DNA forms or the compact formations arranged as concentric circles cannot, in principle, be produced from the helix postulated above by decondensation or flattening. Similarly, they cannot be formed from the straight DNA segments arranged at a certain angle to the long axis of the particle (North and Rich, 1961). Strong evidence in favor of the tertiary-structure suborganization of the intraphage DNA *in situ* is presented by its decondensation patterns.

Analysis of the DNA decondensation products (Rubinstein, 1960; Maestre and Kilkson, 1962; Klimenko *et al.*, 1967; Tikchonenko, 1969) indicates that the tertiary structure of intraphage DNA is most probably formed by lateral aggregates of DNA segments. These aggregates are 7.5–14.0 nm in size and easily decompose into the individual DNA strands when stained by uranyl acetate (Klimenko *et al.*, 1967). The DNA segments are formed by bending the molecule, and the segments of one strand may have various lengths, which easily explains the stepwise decrease of the strand diameter toward its end. For instance, Rubinstein (1960) found the following diameters: 7.5, 5.7, 4.1, and 2.7 nm. These results may be easily interpreted by assuming that a strand includes seven double-stranded segments, one of which is represented by a single double helix and the six others by DNA coils formed by bending the molecule. In this model, DNA coils have various lengths, which provides for a diameter decreasing toward the end of the strand. Formation of such strands might involve the inner proteins, polyamines, and divalent cations.

Thus, DNA packing in the phage head should be some type of circular winding, either in the form of the secondary thread (Klimenko *et al.*, 1967; Tikchonenko, 1969) or in the form of the primary strand (Richards *et al.*, 1973). Only circular winding of DNA may account satisfactorily for the easy transformation of its native tertiary structure into torelike or ring-shaped formations or formations consisting of concentric DNA rings. As discussed above, there are two possible types of DNA packing in the phage head.

The first model represents DNA *in situ* in the form of a sphere or a ball in the isometric virion head or in the form of a prolate ellipse corresponding to icosahedral symmetry with elongated middle region (the T-even phages). In this case DNA is packed in the form of regular or deformed circles whose centers correspond to the virion center and their planes intersect at various angles. The resulting structure looks like a ball of thread, with several layers of rings whose diameters increase from the center to periphery.

According to Richards *et al.* (1973) the second model represents a spool with a coaxially wound individual DNA strand. In this case the planes of the turns should be perpendicular to the particle axis to account for the negative flow birefringence (for the T-even phages). In order to allow for the uniform packing of the phage heads with spool-structured DNA it should be postulated that the number of turns is highest in the center and diminishes toward both ends. We find a few drawbacks in the spool model. Apparently, the spool axis should be a continuous hollow tube passing through the virion center in the plane perpendicular to the virus particle axis. Such a "well" passing through the bulk of the DNA should be very easy to notice on sections, but, in fact, electron microscopy of ultrathin sections does not reveal such a hole in the phage (see below the discussion of the central-cavity problem).

Finally, to account for the DNP negative flow birefringence, as mentioned above, the spool axis should be perpendicular to the phage axis, whereas formation of the circular DNA pattern is possible only if the spool axis assumes the orientation perpendicular to the support film. To save the issue the authors have to postulate reversal of the DNA spool since otherwise, with the parallel arrangement of their axes, DNA would exhibit parallel striations in the micrographs instead of the circular patterns. However, such a DNA pattern has not been observed by these authors or by anybody else. Of course, these negative results cannot by themselves rule out such a packing model. Apparently, the postulated rotation of the spool axis which may occur under the effect of surface-tension forces during drying of the extravirion DNA cannot be carried out inside the capsid. Nevertheless, the micrographs presented by Richards *et al.* (1973) regularly reveal circular DNA packing ***inside*** the heads having only slight damage. Similar structures have also been observed by Manykin (1972) in partially damaged T2 and DDVII phages with uranyl acetate contrasting. All these results taken together make up a formidable argument against the spool model.

Therefore, we shall later consider only the most-realistic hollow-sphere model proceeding from the concept of circular winding of DNA. This model is based on the four known facts: (1) existence of a small central cavity in phages which is similar to the cavity in the RNA-containing isometric viruses (see above), (2) formation of tore or circular patterns in condensed DNA released from phage as the primary product of DNA decondensation, (3) further decomposition of the tore into threads with diameters of 8–15 nm, and (4) flow birefringence and X-ray diffraction data showing some degree of DNA orientation in the T-even phages.

Kilkson and Maestre (1962) and Maestre and Kilkson (1962) were the first to suggest the existence of a central cavity in the T2 phage head. This suggestion was supported by Cole and Langley (1963) on the basis of rather indirect evidence. Later its existence was directly demonstrated by a number of authors by means of electron microscopy of ultrathin sections (Cummings and Wanko, 1963; Cummings *et al.*, 1965; Klimenko *et al.*, 1967). The micrographs published by Cummings and co-workers represent the cavity as a 7 by 15 nm ellipsoid, whereas our results represent it as a sphere with a diameter of about 9 nm. However, some authors using similar techniques have obtained micrographs with no central cavity (Moll, 1963; Margaretten *et al.*, 1966; Simon and Anderson, 1967). Thus Klimenko, Andreev, and Tikchonenko (unpublished data), using different conditions of fixation and dehydration, have studied the factors determining the presence or absence of the central cavity in sections and have found that the cavity disappears with partial denaturation of the intraphage DNA during fixation with formaldehyde and that it is indistinguishable in sections whose thickness exceeds 40 nm. Even though all these results testify strongly to the real existence of such a cavity, there is yet a possibility that the cavity appears as a result of nonuniform contraction during dehydration or embedding of the specimens. However, this seems to be highly unlikely in the light of the data on the easy transformation of the tertiary structure of the intraphage DNA into the toroidal form. When T2 or T4 phages are destroyed under certain conditions [the so-called *p*H–PTA shock or drying in a thin layer of phosphotungstic acid (PTA)] DNA is released into the medium, retaining its condensed form (Klimenko *et al.*, 1967; E. Kellenberger, personal communication;

Richards *et al.*, 1973). Mostly, this DNA form is an ellipse with the major axis of 90–190 nm and a thickness of 18–45 nm. A toroidal structure of similar form may be observed in the semiempty heads of T2 and DDVI phages adsorbed on the isolated cell walls of bacteria (Velikodvorskaya *et al.*, 1968). It is important to note that in this case the authors used direct electron microscopy of the semiempty phage particles after negative contrasting, and, hence, the toroidal DNA structure cannot be an artifact of the ultrathin section preparation technique [for references and discussion see Tikchonenko (1969)].

Nevertheless, DNA in phage cannot have the toroidal form as its dimensions and, in particular, the size of the central hole do not correspond to the size of the central cavity. Similarly, the sections of phage particles lack figures corresponding in size and form to a tore cut along a given plane.

All these data suggest that the tore is the primary product of DNA decondensation, formed either upon going out of the head or directly in the head during ejection of part of the DNA. Therefore, it may be assumed that, generally, DNA in the phage head is an ellipsoid of revolution with the central cavity formed by the secondary DNA thread. This figure may be easily transformed into a tore either during decondensation in the medium or in the capsid when part of the DNA goes out.

The micrographs presented by Richards *et al.* (1973) reveal the phage particles disintegrated or damaged due to drying in the thin layer of PTA. The DNA of such particles, either after release into the medium or in the damaged capsid, is flattened but retains the compact form, although its diameter exceeds by ⅓–¼ the head diameter of the intact particles. The concentric rings of the individual DNA molecules may be easily seen in such flattened DNA bumps, thus demonstrating the easy transformation of the densely packed DNA into a packet formed by DNA rings. It may be suggested that the next step of decondensation is the transformation of this shape into a tore, though no experimental evidence of this has yet been obtained.

Although there are two possible models for the transformation of the native structure of the DNA into a tore or into concentric circles (a rigidly fixed one and a mobile one) [see Tikchonenko (1969)], clearly the second model is preferable. According to the mobile model, DNA transformation implies relative sliding of the adjacent DNA segments. This segment sliding apparently takes place in the head during injection of the phage DNA into the cell and *in vitro* during DNA decondensation. The high speeds of both processes imply that the postulated sliding does not necessitate overcoming high potential energy barriers due to adhesion forces between adjacent DNA turns. The most likely governing factor of this process is the free-energy gain in the conversion from the closely packed DNP condensate to a loose-coil state. It is important to note here that the similarity of the tores formed *in vitro* during the *p*H–PTA shock and *in vivo* in the partially empty heads suggests that there are forces acting in the native DNA condensate *in situ* which potentially tend to convert the tertiary structure into a tore as the first stage of decondensation. In this connection it should be noted that the ring toroidal DNA structures similar to those described above are easily formed during aggregation of the deproteinized DNA under certain conditions. It may be suggested, therefore, that under the conditions favoring DNA aggregation the appearance of the toroidal form is a natural, energetically favored process reflecting the inherent properties of the double-stranded polymer and its tendency to form a certain tertiary structure.

In the section concerned with the ORD and CD spectra of the intraphage DNA (see below) we shall discuss in more detail the attempts to simulate *in vitro* those conditions under which DNA may exist in DNP of various types. In connection with the above toroidal structures observed with disintegration of various phages, we should pay attention to their similarity to the so-called compact DNA forms. These forms have been observed by a number of authors in DNA interactions with polycations and alkaline proteins (Shapiro *et al.*, 1969; Inoue and Ando, 1970; Olins and Olins, 1971; Shih and Fasman, 1971; Gottesfeld *et al.*, 1972; Suwalsky and Traub, 1972), with various synthetic water-soluble polymers (Lerman, 1970; Jordan *et al.*, 1972; Evdokimov *et al.*, 1972, 1973, 1974*a,b*; Akimenko *et al.*, 1973; Varshavsky *et al.*, 1973), and also with acidification (Dore *et al.*, 1972, 1973). Some of these compact DNAs obtained in the presence of polylysine and various histones (Shapiro *et al.*, 1969; Haynes *et al.*, 1970; Olins and Olins, 1971) and polyethylene glycol (Jordan *et al.*, 1972; Evdokimov *et al.*, 1972, 1973) have been shown by electron microscopy to have the form of a tore with a diameter of 60–170 nm. It is assumed that the direct cause of DNA compactization and tore formation is dehydration of DNA. According to Evdokimov *et al.* (1974) in the presence of polyethylene glycol there occurs a loss of water localized in the large and small troughs of the DNA without decrease of hydration of the DNA phosphate groups. At acidic *p*H values the DNA takes on the form of a sphere without the central hole (Dore *et al.*, 1973).

This leads to the natural question as to what extent this exciting similarity of the electron microscopic images relates to the actual similarity of the DNA compact structures *in situ* and *in vitro*. Evidently, the extent of similarity should not be overestimated for the simple reason that DNA *in situ* cannot be packed in the form of a tore (see above). In this case the tore is a transient formation at the first decondensation stage which starts after DNA release from the virion and ends by assuming the normal tertiary structure of the random coil.

The sequence of the decondensation stages occurring with phage disintegration (I) and the sequence of events with the reverse process of DNA condensation *in vitro* (II) may be written in the following form:

(I) DNA *in situ* → Tore → 7.5- to 14.0-nm Thread → Coil

(II) ← Tore ← ? Strand ? ← Coil

Apparently, the condition determined in the experiments *in vitro* simulate the factors governing one of the transient decondensation stages of the phage DNA in solution rather than reconstruct the conditions prevailing in the phage particle or in the infected cell during the filling of the phage head with DNA. When any of the above factors are present, this transient stage is somehow preserved and the tore can exist indefinitely.

2.2. The Secondary Structure of DNA

When we discussed the peculiarities of DNA packing in phages and, particularly, the packing density we already assumed that the secondary structure of DNA might be altered to a certain extent. Various authors have repeatedly suggested denaturation alterations of the intraphage DNA due to either its over-dense packing or numerous molecular bends (Dunn and Smith, 1958; Bonhoeffer and Schachman, 1960; Thomas, 1963; Bendich and Rosenkranz, 1963), but the first attempted experimental verification of this suggestion yielded negative results (Bonhoeffer and Schachman, 1960). Only later studies have shown that [for references see Tikchonenko (1969)] the secondary structure of the intraphage DNA is indeed altered, although these alterations are not a trivial disturbance of the ordered secondary structure (similar to free DNA denaturation in solution) but a complex, reversible, conformational transition due mainly to the interaction between DNA and protein. These conformational alterations of the intraviral DNA are evidenced by changes of both its optical properties and base reactivities.

2.2.1. DNA Structural Heterogeneity and Conformation of Its Double-Stranded Parts

It is highly probable that in the intraphage DNA there occur not conformational alterations of the molecule as a whole, but only local alterations. In the phage virions, in addition to the DNA regions with the typical Watson–Crick double-helix structure, there apparently also exist parts with altered secondary structure (Tikchonenko *et al.*, 1966; Tikchonenko, 1969). Generally, such heterogeneity is associated with the complexity of the virion structural organization, the presence of various proteins, and the close contact between them and part of the DNA, as well as with the above peculiarities of the DNA tertiary structure *in situ* characterized by numerous bends, lateral aggregation, etc.

Practically all the results of the molecular biology studies of bacteriophages of this type indicate that the greater part of the DNA in Sd, T2, T7, and similar phages has the perfect double-helix structure *in situ*. Therefore, numerous experiments are aimed not at substantiating the existence of regions with the regular double-helix structure in the phage but at finding regions with altered secondary structure. The presence of regions of both types in intraphage DNA is shown by UV absorption spectra and by infrared spectra, which for the sake of continuity will be discussed later, as well as by the results obtained with dyes, e.g., acridine orange, proflavine, pinocyanol (PNC), and others. Staining of all the acridine-orange-permeable phages gives rise to two luminescence maxima: one at 530 nm, typical of the monomeric-dye form interacting with double-stranded DNA, and one at 640 nm, typical for the dimeric dye forming complexes with the denaturated DNA (Gabrilovich *et al.*, 1968, 1970; Basu, 1971; Pysarevsky *et al.*, 1968; Kuriatkowski *et al.*, 1973; Dusenbery and Uretz, 1972). Dye complexes with the free DNA of these phages naturally give rise only to the green luminescence typical of the double-helix structures.

Only a limited amount of intraphage DNA is accessible for acridine dyes, not more than 60% in the case of $T4_{O_1}$ phage (Dusenbery and Uretz, 1972) and 50% in the case of Sd phage (Permagorov *et al.*, 1969). Basu (1971) has noted that interaction between acridine orange and DNA in phages seems to be of two types, and the predominance of one or the other of the types is determined by the temperature. Though the mechanism of these interactions has not yet been studied in detail it has been found that the interaction between the dye and the intraphage DNA intensifies sharply at 50–52°C. The effects observed are not associated with the partial disintegration of the phage and the DNA release into solution since they are absent in the concurrent experiments with ethidium bromide. It has been suggested that at room temperature the dye interacts with that part of the intraphage DNA which has the altered secondary structure and is localized at the periphery of the virion. At the elevated temperatures the higher lability of the intraphage DNA structure allows the acridine orange to gain access to the inner parts occupied by DNA. Interestingly, at these temperatures the contribution of the light scattering to the optical density at 260 nm changes significantly (Basu, 1971).

Dusenbery and Uretz (1972) believe that the two types of interaction are the strong binding of dye to DNA due to intercalation and the weak cooperative binding leading to dye stacking. They have found that strong interaction is greatly reduced in the intact phage, the cooperative weak binding being dominant.

Similar results have been obtained in the experiments with PNC, which, in contrast to acridine orange, is dimerized at the double-stranded parts of DNA. The dye dimers induce characteristic optical rotatory dispersion with a maximum at 560 nm (Permagorov *et al.*, 1969). In the Sd phage only half of all DNA in the virion is accessible to PNC and the rest of the DNA does not react with dye under these conditions. In this case only 30% of the DNA–dye complexes (i.e., 15% of the total amount of the nucleic acid) demonstrates induced optical activity, evidencing their double-helix structure. The rest, 70% of the DNA–dye complexes (i.e., 35% of the total DNA in the virion), is incapable of PNC dimerization. As disruption of the ordered secondary structure of DNA in the experiments *in vitro* impedes PNC dimerization but does not affect binding of the dye, it may be suggested that the parts of the intraphage DNA binding PNC but incapable of providing for its dimerization also lack the ordered base stacking. Theoretically, it is possible that the failure of PNC to dimerize may also be due to some additional spatial hindrance *in situ* (Permagorov *et al.*, 1969).

These results may be subject to different interpretations. Most of the above authors proceeded from an analogy between the experiments *in vitro* and *in situ* and believe that different abilities of free and intraphage DNA to yield complexes with dyes may be understood in terms of the different amounts of the ordered secondary structure. On the other hand, Dusenbery and Uretz (1972) pointed out that nearly all of the studies using dyes were performed in relatively dilute solutions, while nucleic acids *in situ* are mainly densely packed. Therefore, there theoretically exists a possibility that the interaction of dyes with DNA in gels and in phages may be influenced by the different hydration of DNA under these conditions. The above viewpoint is strengthened to some extent by the fact that the same "red shift," reflecting a change in weak cooperative interactions, is observed in dye complexes in the $T4o_1$ phage and in DNA gels under 90–65% relative humidity. Dusenbery and Uretz (1972) believe that the main reason for the change in the fluorescence color in the phage is a difference in DNA hydration. Obviously, the dense packing of DNA *per se* cannot influence the dye complexing mechanism, as the fluorescence of dye complexes with DNA gels at relative humidities greater than 90% is the same as in dilute DNA solutions. These authors do not report whether the change in hydration influences also the strong dye binding to DNA.

Thus, the DNA molecule in various phages apparently consists of regions with different secondary structures. This conclusion is substantiated by calorimetric measurements (Semenov *et al.*, 1971). There are divergent views concerning the state of the double-helix regions of the intraphage DNA: some authors believe that these regions have the B configuration typical of the free double helix in solution, while others are in favor of the C configuration. The existence of the B configuration is confirmed by the fact that the X-ray diffraction patterns for the oriented preparations of some phages have a discernible reflection at 0.34 nm (North and Rich, 1961; Dembo *et al.*, 1965). The latter authors, proceeding from the results of small-angle X-ray scattering for nonoriented preparations of large and medium-sized phages, arrived at the same conclusion (Dembo *et al.*, 1965). The C configuration of DNA should give rise to a reflection at 0.33 nm; in fact, this was observed by Katz and Rich (1966) but was considered to be due to water. Naturally, these contradictory ten-year-old results obtained by different authors urgently require a reinvestigation before being used as strong arguments in a scientific discussion.

The infrared spectra of bacteriophages (Shie *et al.*, 1970, 1972*a*) suggest that the double-helix regions of the intraphage DNA have the B configuration. The technical potentialities of spectrophotometers and the absorption characteristics of nucleic acids and proteins in the infrared range use to this end [while working with the intact virus (!)] only the absorption bands due to the sugar–phosphate skeleton of DNA. The theoretical calculations and the experiments with model compounds have demonstrated that the bands at 1230 cm^{-1} and 1055 cm^{-1} are rather sensitive to the DNA conformational states. The first band is due to the antisymmetric stretching vibration of the

$$\rangle PO_2^-$$

group and the second is due to the conjugated vibration of the C^5–O^4 bond of the sugar–phosphate skeleton (Sutherland and Tsuboi, 1957;

Shie *et al.*, 1972*a,b*). In the conformational transition of the A ⇆ B type the variation of the frequency of the 1230 cm^{-1} band is due to alteration of the hydration of the phosphate groups, and variation of the intensity of the 1055 cm^{-1} band reflects the structural state of the sugar–phosphate DNA skeleton. The infrared spectra of the free and intraphage DNA have proved to be the same within the measurement accuracy both in films and in solutions (Shie *et al.*, 1970, 1972*a*). Heating of the phage suspension leading to capsid rupture and DNA release into solution does not alter the above spectral parameters associated with the 1230 and 1055 cm^{-1} bands. These results suggest that the intraphage DNA, or more exactly its sugar–phosphate skeleton, has the B configuration typical of double-helix DNA in solution. It should be noted that hydration of the DNA phosphate groups *in situ* and *in vitro* is the same, undermining to some extent the assumptions proceeding from the theory of dehydration as a factor responsible for conformational changes of DNA molecules. Nevertheless, the separate loss of water in the trough of the molecule without the concomitant dehydration of phosphates cannot be excluded (see above about the tertiary structure of intraphage DNA).

The behavior of the bands at 1055 and 1230 cm^{-1} indicates that the intraphage DNA undergoes a typical A ⇆ B transition, with alteration of the relative humidity of the medium evidencing the absence of rigid DNA fixation in the phage and its conformational mobility.

Here a natural question may arise as to how accurately we can evaluate from these data the amount of DNA with a given secondary structure. If we have in mind the ordered secondary structure and its denaturation alterations, such evaluation may be done proceeding from the parameter D_{1055}/D_{1088}, which characterizes the intensity of the band at 1055 cm^{-1} and is 1.11 and 0.85 in native and completely denaturated DNA, respectively. In other words, the range of variation of this parameter is about 30% of its value for native DNA, and, as the mean error is about 3%, this is close to the accuracy of determination of the spiralization percentage for nucleic acids proceeding from hyperchromism measurements. It may be suggested that either the Sd phage lacks the denaturated DNA regions completely or their amount does not exceed 3–5% of the total nucleic acid content in that virion. Thus, those alterations of the optical properties and reactivities of the bases which are typical of intraphage DNA (see below) are not associated with the trivial denaturation disturbance of the sugar–phosphate skeleton *in situ*.

Earlier, a number of authors studying melting of intraphage DNA in the presence of formaldehyde (Tikchonenko and Dobrov, 1969; Inners and Bendet, 1969) suggested a certain superstabilization of the double-helix regions in Sd and T2 phages. The latter studies have shown that this result is an artifact which is apparently due to the cross-links between DNA and protein introduced by formaldehyde (Dobrov *et al.*, 1972*a*; Andriashvili *et al.*, 1972). These authors have found a solvent (0.0015 M $NaNO_3$) which increases the thermal stability of the Sd phage capsid and permits melting intraphage DNA without disintegrating the virions. The melting temperature (T_{mel}) of the intraphage DNA (of its double-helix regions) has proved to exceed by only 4°C the T_{mel} of the deproteinized DNA, and these four extra degrees are due to the thermostabilizing effect of the protein rather than to the structural features of DNA *in situ*. Similar to free DNA, the T_{mel} of the intraphage DNA drops with increasing ionic strength, amounting to 70 and 82°C, respectively, in 0.0015 M and 0.1 M $NaNO_3$. Melting in these experiments is conducted, naturally, in the presence of 1.5% formaldehyde to prevent phage disintegration in high-ionic-strength media.

Maestre has somewhat different views concerning the secondary structure of the intraphage DNA; he believes that the above processes involve not the double-helix regions of DNA but the molecule as a whole, which has a special conformation *in situ* (Maestre and Tinoco, 1967; Tunis-Schneider and Maestre, 1970). This approach is based on the studies of ORD and CD spectra of intraphage DNA.

The ORD and CD spectra for intraphage DNA are very peculiar (Maestre and Tinoco, 1967; Gorin *et al.*, 1967; Maestre, 1970; Tunis-Schneider and Maestre, 1970; Tikchonenko *et al.*, 1971; Holzwarth *et al.*, 1974). Despite the considerable individual differences between the ORD and CD spectra for various phages, their common feature is a sharp decrease of the amplitude of the peak of positive rotation or of the CD band, respectively, the red shift of the transition point, as well as the optical activity in the 300–350 nm range, where there is no intrinsic absorption due to the phage components. Under mild conditions, disintegration of phages and DNA release are accompanied by the immediate restoration of the optical activity pattern typical of double-helix DNA, or more exactly, of its B configuration (Tunis-Schneider and Maestre, 1970). The differences in the ORD and CD spectra between the free and intraphage DNAs significantly exceed the change in the optical properties observed upon denaturation of the double-helix DNA (Maestre and Tinoco, 1967); therefore, the ORD and CD patterns observed *in situ* cannot be explained as a result of trivial disturbance of the base stacking and complementary interactions.

The fascinating variety of the CD spectra yielded *in vitro* by different forms, conformations, and complexes of DNA once more emphasizes the difficulties encountered in interpreting the observations made on phages. The CD spectra of the intraphage DNA bear a certain similarity as regards the form and intensity of the bands to the CD spectra of DNA in the presence of ethylene glycol (Nelson and Johnson, 1970) or in concentrated neutral salt solutions (Tunis-Schneider and Maestre, 1970; Permagorov *et al.*, 1970). The above conditions may, to a certain extent, influence the water activity and, finally, DNA hydration, though the base rotation angle may also vary with the ionic strength. However, the CD spectra for the intraphage DNA greatly differ in band position and intensity from the spectra recorded for DNA complexes with poly-L-lysine (Cohen and Kidson, 1968; Shapiro *et al.*, 1969; Haynes *et al.*, 1970; Chang *et al.* 1973) and for ψ condensates of DNA obtained in the presence of polyethylene oxide or polyethylene glycol and NaCl (Lerman, 1970; Lerman *et al.* 1969; Jordan *et al.*, 1972; Varshavsky *et al.*, 1973; Evdokimov *et al.*, 1973, 1974). Closely similar are the results for the DNA complexes with f1 histone, although the intensity of the CD bands in this case is greatly reduced (Fasman *et al.*, 1970). However, the DNA complexes with f2 histone yield entirely different CD spectra, resembling those for RNA (Shih and Fasman, 1971). There is no generally accepted explanation of the differences observed in the DNA optical activity, but it seems to be highly unlikely that there may be the same cause in all the cases discussed. Therefore, an explanation of the structural peculiarities of the intraphage DNA proceeding from similarity or differences of the CD spectra would be highly questionable.

The main problem in the application of the circular-dichroism method for investigation of any particulate systems, including viruses, is the presence of pronounced distortions in their spectra. One must treat as "particulate" such systems as viruses, bacteria, chloroplasts, ribosomes, and so on. In all the above cases (interaction with proteins, high ionic strength, condensed form, etc.) the DNA structure becomes more compact, resulting in sharply increased light scattering. Two sources of distortion of CD spectra in particulate light-scattering systems have been recognized [for references see Philipson and Sauer (1973)]. The first one is a flattening effect which has its origin in "absorption statistics." According to Philipson and Sauer (1973), "the suspension will have regions (the particles themselves) where absorption is very high and other regions (between suspended particles) where absorption is negligible." Flattening effects on optical activity have been reviewed by Gordon and Holzwarth (1971). Since this effect is inherent in the particulate materials, it cannot be corrected instrumentally.

The second source of distortion is light scattering. Any particulate system, including viruses, may produce both the classical Raleigh–Mie scattering due to the isotropic structure, and the anomalous types of scattering due to the anisotropic properties. The most important is the second kind of light scattering. The studies done by many authors with the CD of different particulate systems (chloroplasts, phages, algae, etc.) have shown that the observed CD spectra are critically dependent on the proportion of scattered light. This dependence indicates that different quantities of left and right circularly polarized light are scattered from the measuring beam of the instrument. It is noteworthy that it is not particulateness *per se* which is responsible for the large differential scattering; Philipson and Sauer (1973) and Holzwarth *et al.* (1974) believe it is rather the scattering due to the complex ordered or stacked structures. This possibility has been brilliantly demonstrated in the case of the CD spectra of chloroplasts, the peculiarities of which have proved to be due to the peculiar arrangement of chlorophyll within the chloroplast membrane (grana). The drastic changes in the CD spectra of chloroplasts after their disruption are not due to the changes in light scattering *per se,* but mostly to the dispersion of grana structure. Most convincing is the fact that the turbid suspension of blue-green algae, which lack the ordered organization of their pigments in the grana form, has the same intensity of light scattering as chloroplasts, but the CD spectra (corrected) of algae are similar not to CD spectra of intact chloroplasts but to the spectra of the disrupted ones. Thus, it is highly probable that the different scattering of left and right circularly polarized light is a function of the structure of the scattering particles.

Of course, the Mie scattering may also influence to some extent the CD spectra of particulate systems. Attempts have been made by Gordon (1972), Gordon and Holzwarth (1971), and Holzwarth *et al.* (1974) to account for Mie scattering contributions to CD spectra. The theory predicts the red shift and distortion in ORD and CD spectra for particles composed of DNA with a radius (R) of 50–100 nm, density of $\sim$1 g/ml, and mean refractive index of 1.20. In particular, the value of this shift is dependent on particle dimensions: for R = 50 nm the shift is 3 nm, for R = 100 nm it is $\sim$10 nm (Gordon, 1972). In the first approximation the calculated spectral distortions resemble the real ones observed in particulate systems.

Especially interesting is the attempt to do such calculations for bacteriophages (Holzwarth *et al.*, 1974), although some authors consider such attempts as "sophisticated ones, because they are done proceeding from symmetrical model systems such as a solid sphere" (Philipson and Sauer, 1973). They agree with Dorman and Maestre

(1973) that the large effects on the absorption and CD spectra may result from differential scattering due to an asymmetrically ordered structure (see above).

Dorman and Maestre (1973) have used the fluorscat method to allow for light scattering and found that the CD spectra registered in the conventional way, and the CD spectra corrected for light scattering, do not differ greatly, and that the difference mainly concerns the band intensity. The CD spectrum registered by the fluorscat device for the intraphage DNA had the band intensities similar to those for the DNA C form, but the C form lacked the red shift. However, the calculations of Holzwarth *et al.* (1974) demonstrated, in accordance with the above, that the phage CD spectra are mainly influenced not by the Mie scattering but by the anisotropy of the intraphage DNA. These authors represent the virus particle as a perfect sphere 100 nm in diameter with a protein shell 7.5 nm thick and the diameter of the region occupied by DNA of 85 nm (the tail is ignored in such calculations). The calculated CD spectra for the isotropic spherical phage model had a larger amplitude of the positive CD band at 280 nm than the measured CD spectra and no optical activity in the long-wavelength region of the spectrum (300–350 nm), although the calculated signs and the shape of the bands resembled those of the measured spectra. Proceeding from this, Holzwarth *et al.* (1974) have suggested that the densely packed intraphage DNA may actually possess *anisotropic* properties. In this case the structural anisotropy may be evidenced by different absorption or scattering of the left- and right-polarized light leading to the anomalies in the CD and ORD spectra. Thus, according to these authors, the profound difference in the CD spectra of the intact and disrupted phages arises from a combination of conformational change and supramolecular ordering of densely packed DNA. Meanwhile, it should be remembered that the same authors have satisfactorily explained the intraphage DNA hyperchromism proceeding from the isotopic model and the Mie scattering.

As such geometrically different formations as the intraphage DNA, tissue DNPs, and various condensed or partially dehydrated DNA forms have similar CD spectra, this similarity should apparently be based on some essential common features of the structural organization as a whole. Holzwarth *et al.* (1974) believe this to be due to the anisotropy of properties in all the liquid-crystalline structures, irrespective of the presence or absence in them of DNA. Owing to this similarity of the structural organization, such diverse objects as the chloroplast membranes and the intraphage DNA possess similar CD spectral characteristics. The above authors have found three similar characteristics of the CD spectra due to anisotropic packing of components in the liquid-crystalline systems.

1. The CD amplitudes are considerably greater than those intrinsic to their molecular constituents. Thus, the intensity of the CD spectra for intact chloroplast membranes (Holzwarth *et al.*, 1974), DNA complexes with poly-L-lysine (Shapiro *et al.*, 1969), or the ψ^+ condensate of DNA (Jordan *et al.*, 1972; Evdokimov *et al.*, 1973, 1974) is indeed greater by an order of magnitude than the intensity of the CD spectra for their constituents or for control DNA. However, the differences between the CD spectral intensities are rather small for the intraphage DNA and for a number of DNA complexes [for references see Jordan *et al.* (1972)].

2. CD is observed in spectral regions where there is no absorption by the components of a given system. In particular, the optical activity displayed by intraphage DNA in the 300–350 nm range cannot be ascribed either to the intrinsic CD of any particular DNA secondary structure or to the Mie scattering.

3. The CD bands of all the liquid-crystalline systems resemble in form the absorption spectra of their components but have a hook of opposite sign.

Thus, the anisotropy of the liquid-crystalline structure and the resulting anomalous scattering seem to provide a possible, though not yet proved, explanation of the CD and ORD anomalies of intraphage DNA. The second and the third approaches treat the CD and ORD anomalies of intraphage DNA as a result of structural alterations which DNA undergoes in phage. According to the second approach, the peculiarity of the intraphage DNA optical activity is due to its special conformation; the C form according to the results of Maestre's group (Tunis-Schneider and Maestre, 1970; Maestre, 1970; Holzwarth *et al.*, 1974) and a particular form according to Lerman (1970) and Lerman *et al.* (1969). In reality (see above), the peculiarities of the CD spectra of phages and other particulate systems may result from any combination of the above-discussed causes.

Maestre's approach is the most developed one. The model experiments with DNA films and solutions have shown that where DNA is in the C form as demonstrated by the X-ray diffraction results, the pattern of the optical activity changes, becoming similar to that for the phages. At the same time, in some studies [for example, Brunner and Maestre (1974)] concerning ψ spectra of polynucleotide films the authors have been inclined to explain the above anomalies of CD spectra exclusively by the liquid-crystal properties of these systems. However, the similarity of the spectra in itself cannot be treated as unambiguous and exhaustive proof of identity of the structures. Moreover, the X-ray diffraction results show that in tissue and phage nucleoproteins DNA does not have the C form, but the B form, or, more exactly, one of its configurations belonging to the B-form family (Haynes *et al.*, 1970; Shapiro *et al.*, 1969). DNA also appears to have the same B form in the ψ complex (Jordan *et al.*, 1972).

However, while making the Mie calculations for the CD spectra of the intraphage DNA, Holzwarth *et al.* (1974) proceed from the refractive index typical of the B, but not C, form of DNA. Substitution into such calculations of the parameters typical of the C form of DNA produced less agreement between the experimental and theoretical curves. In this case the calculated CD spectra of the intraphage DNA lacked, in particular, the positive CD band at 290 nm. Maestre (personal communication) explains this by the inability of the Mie scattering theory to account for the CD properties of viruses. In his opinion "it is possible to have C geometry and anisotropic long-range ordering in the T2 DNA *in situ*." However, for the C configuration in the phage head one must pay with "less resemblance" between calculated and measured CD spectra, thus depreciating to some extent the conclusions made in Holzwarth's paper.

According to Tunis-Schneider and Maestre (1970), one of the main factors of DNA conversion into the C form is the partial dehydration. In model experiments the CD spectra of DNA films at relative humidities below 60% became similar to the CD spectra in 4–6 M NaCl or LiCl. Dehydration is also a natural explanation for the anomalies of CD spectra in the DNP or artificial DNA complexes with basic polypeptides as well as for DNA in salt solution. Polycations or metal ions neutralize charges on phosphate groups and exclude water molecules from the phosphate vicinity and other moieties of double-stranded DNA (Tunis and Hearst, 1969; Tunis-Schneider and Maestre, 1970; Chang *et al.*, 1973).

However, there exist some doubts concerning such strong DNA dehydration in the phage. The infrared-spectrophotometry results indicate that the hydration of the phosphate groups in the free and intraphage DNA is the same (Shie *et al.*, 1970). Although it is possible that dehydration takes place at the expense of the bases and does not involve the phosphate groups (see above), this should be especially substantiated since DNA dehydration *in vitro* at the start causes the phosphate groups to lose water (Falk *et al.*, 1962). We believe that structural heterogeneity of DNA exists in the phage irrespective of which conformation is present in the virion (B, C, ψ, or some intermediate form).

The similar point of view concerning tissue DNP and the complexes of DNA with basic polypeptides has been developed by Spitkowski *et al.* (1969) and Chang *et al.* (1973). Without consideration of the structural heterogeneity of intraphage DNA it is difficult to account for a number of experimental facts discussed below.

Even though at present there are, unfortunately, no comparative data on the CD spectra of various phages, such data are available for the ORD spectra (Maestre and Tinoco, 1967; Tikchonenko *et al.*, 1974*a,b*). Various phages have been proved to have significantly different amplitudes of the peak at the 270–280 nm range, the value being either slightly positive or slightly negative. For instance, the value of $[\alpha]_{268}$ for DNA in the phages Sd and T2 is +500° and −1100°, respectively, i.e., the difference in the absolute magnitude is 1600°. Since in this spectral range the protein components of both phages have almost no influence on the $[\alpha]_{268}$ values (Maestre and Tinoco, 1967), we should naturally be concerned with the factors determining such considerable interspecies differences in the peak amplitudes if both DNAs have *in situ* the same form (C, or anything else).

This inconsistency is eliminated by the third approach, i.e., assuming that there are intraphage DNA regions differing in the secondary structure and interaction with protein and, hence, in the character of their ORD and CD. In this case the spectra of the intact phages are formed by averaging the ORD and CD spectra for each of the structural elements. We believe that the double-helix regions of the intraphage DNA may have *in situ* the normal ORD spectra with $[\alpha]_{268}$ being equal to +2300°, whereas the regions of altered conformation may exhibit a strong negative rotation in this spectral range of about −3000° (Tikchonenko, 1974*a*; Tikchonenko *et al.*, 1974*a,b*).

Concerning the mechanisms inducing the strong negative rotation in these parts of intraphage DNA, it is difficult to state anything definite. The three following possibilities may be considered in the general form as such mechanisms: altering the stacking angles θ between the bases, the mixed stacking of bases and amino acids, and rotation of bases around the *N*-glucoside bond, the extreme case of which is an *anti* → *syn* transition. The first cause seems to be more popular as it agrees with the recent CD theories (Johnson and Tinoco, 1973; Moore and Wagner, 1974). The second cause has been considered in more detail above (see the section on rodlike viruses). The third approach is much less developed and has been used only by Michelson *et al.* (1970) for explanation of the negative rotation displayed by polybromoguanylic acid and by McGavin *et al.* (1966) and McGavin (1971) for constructing the four-stranded DNA models. This factor has also been suggested to be one of the main causes of the difference between conservative and nonconservative CD spectra (Moore and Wagner, 1973).

The third approach has been used by Tikchonenko (1974*b*) in his attempt to explain the whole set of data concerning the intraphage DNA. We should stress two facts concerning this assumption. First, the terms *syn* and *anti* in this case are used in close analogy since in the B form of DNA the angle ϕC is $-86°$, whereas for the mononucleotides in the anti conformation this angle varies from $-75°$ to $+15°$ and for the mononucleotides in the *syn* conformation it varies from $+105°$ to $+195°$ (Sundaralingham, 1969). Thus, in the B form of DNA the position of the bases with respect to the sugar residues may be described in terms of the *anti* conformation only by somewhat stretching the argument (Yang and Samejima, 1969). Nevertheless, manipulations with a stereomodel of the double-stranded DNA indicate that with rotation about the *N*-glycoside bond of the *anti* $\rightleftarrows$ *syn*-transition type the bases are removed from the stack without disturbing the sugar–phosphate skeleton. It is quite probable that to obtain the effects desired (disruption of base stacking, alteration of the reactivities of the functional groups as well as of the absorption, CD, and ORD spectra) there is required not the complete *anti* $\rightleftarrows$ *syn* transition but lesser changes of the angle ϕC.

The interpretation of the intraphage CD spectra in terms of release of the bases from usual stacking does not contradict the optical activity theory developed by Tinoco, although it certainly raises some complicated problems. Of course, disordered shifting of bases outside of the stacking geometry, as it takes place with thermal denaturation, would reduce all the magnitudes of the CD bands. At the same time, the situation must be different if one considers the *anti* $\rightleftarrows$ *syn* transition of some *adjacent* bases, i.e., the "ordered destacking" of large regions of intraphage DNA. It may be of interest to consider the consequences of this suggestion from the standpoint of current optical theories.

Second, the *anti* $\rightleftarrows$ *syn* transition is well known, both theoretically and experimentally, only at the monomer level and little is known about such transitions in polymers. The only experimental study in this field has been done by Michelson *et al.* (1970) with poly-8-bromoguanylic acid, which the authors believe to have not the usual *anti*, but the *syn*, conformation of the nucleoside residues. To this fact the authors ascribe the negative optical rotatory dispersion in the 260–300 nm range revealed by this polymer but not by unmodified polyguanylic acid. The *anti* $\rightleftarrows$ *syn* transition is caused by the bromine atom limiting the rotation of the base around the *N*-glycoside bond. Arguing this standpoint, Maestre (personal communication) indicates that the CD spectrum of brominated polyguanylic acid is negative in the 280–300 nm region but has a broad positive peak in the 240–260 nm region which is absent in the intraphage DNA. However, one does not have to assume the total identity between intraphage DNA and bromoguanylic acid CD spectra. The negative rotation is believed to be characteristic of the *syn* conformation of a nucleoside residue (Michelson, 1963; Kochetkov *et al.*, 1970).

Such a transition in the intraphage DNA may be due to the interaction between the bases and the protein. Manipulations of the DNA stereomodel indicate that the base rotation around the *N*-glycoside bond may be accomplished without disturbance of the phosphodiester backbone of the molecule, which explains the lack of significant alterations in the state of the phosphate groups of the intraphage DNA as compared to the free DNA (Shie *et al.*, 1972*b*).

If we assume that two-thirds of the DNA in the Sd phage has a double-helix structure with $[\alpha]_{268}$ typical of the B configuration of DNA in solution (+2300°), then the third of the intraphage DNA which possesses a special conformation should have $[\alpha]_{268}$ equal to −3000°, considering that the intact phage rotation is +500°. The model experiments indicate that the natural and artificial polynucleotides may have such negative rotation values under certain conditions (Davidson and Fasman, 1969; Fasman *et al.*, 1970; Nelson and Johnson, 1970; Green and Mahler, 1970; Michelson *et al.*, 1970; Travers *et al.*, 1970). Naturally, this explanation seems to be rather hypothetical and requires rigorous experimental verification, but it accounts satisfactorily for a number of diversified experimental facts in terms of a unified concept (see below).

The stumbling block of the second and third approaches is the observed optical activity in the long-wavelength region of the CD and ORD spectra, where there is no absorption by either DNA or protein. It may be, however, that the above three characteristics of the CD spectra of the liquid-crystalline structures are, in fact, a random assortment of properties. In particular, it may be assumed that the second and third characteristics are, indeed, due to the anisotropy of the liquid-crystalline structures, whereas the first characteristic is determined by the specific structural organization of a given system. This assumption is to a certain extent confirmed by the discrepancy between the differences existing between the CD of the intact system and its individual components observed in the phages and chloroplasts and, also, by the strong variability of the positive CD band amplitude in the 280–290 nm range (see above). It is appropriate to remember here that CD in this spectral range is extremely sensitive to the slightest conformational alterations of the DNA structure and, in particular, to the base-tilt angle. It is just this factor that explains the sharp differences between the CD spectra for the B and C configurations of DNA or for DNA in high- and low-ionic-strength media (Permagorov *et al.*, 1970).

2.2.2. Disturbance of Base Stacking and Complementary Interaction

As mentioned above, disturbance of usual base stacking and complementary interactions in a part of the intraphage DNA is indicated by both the optical properties of the DNA *in situ* and the sensitivity of its functional groups to certain chemical agents. The first to suggest this were Tikchonenko *et al.* (1966), who analyzed the UV absorption spectra of the intraphage DNA. However, this suggestion had to be supported by the results of other independent methods since the optical studies of viruses generated a number of theoretical and procedural complications.

2.2.2a. UV Absorption

The greatest stumbling blocks for the correct application of the optical methods to study intraviral nucleic acids are allowing for light scattering and determining the protein contribution. Both problems were discussed in detail in an earlier review (Tikchonenko, 1969) which is recommended to readers interested in this field. Here we shall only mention that at present the following methods are generally used to account for light scattering and to determine the true absorption of the viral particles: the well-known extrapolation method, the integrating-sphere method (Rvachev *et al.*, 1968; Gratton, 1971), and the fluorscat method (Dorman and Maestre, 1973). The other techniques occasionally used for the determination of light scattering are discussed in the above review.

Because of light scattering (σ), the measured absorption of solutions of particles or macromolecules (A) is known to be greater than the true absorption (A_T). Therefore, $A_T = A - \sigma$. The extinction due to the Rayleigh–Mie scattering is given by $\sigma = K/\lambda^n$, where λ is the wavelength and K is a parameter determined by the solute concentration, its molecular weight, and the refractive index of the solvent. The value of σ in the virus' intrinsic absorption range is determined by extrapolating the plot of log OD *vs.* log λ from the spectral range where there is no absorption of the virus components (usually 320–370 nm) to the virus intrinsic absorption range (Doty and Steiner, 1950; Schauenstein and Bayzer, 1955; Englander and Epstein, 1957; van de Hulst, 1957; Basu and Das Gupta, 1967). For isometric isotropic particles of constant polarizability whose size is small compared to the wavelength of the scattered light, this plot is theoretically predicted to be a straight line with a slope (tan α) of -4. For highly elongated, thin particles tan α should be -3 and for large particles, -2 (van de Hulst, 1957). These relationships are strictly valid in the spectral regions where the particles do not absorb. Theoretically, the scattering in the intrinsic absorption range may be anomalous and may differ greatly from the light scattering in the spectral ranges where there is no absorption by the particles. Therefore, some authors consider the extrapolation procedure to be unsuitable for determining A_T (Eisinger, 1966; Olins *et al.*, 1967). However, numerous experimental observations and some theoretical calculations indicate the absence of anomalous scattering by viruses in the range of intrinsic absorption by protein and nucleic acids (Schauenstein and Bayzer, 1955; Englander and Epstein, 1957; Leach and Sheraga, 1960; Tikchonenko *et al.*, 1966, 1974*a*; Tikchonenko, 1969; Sinha and Misra, 1971; Inners and Bendet, 1969). The validity of the extrapolation procedure is confirmed also by spectrophotometric measurements with the integrating sphere either specially designed (Rvachev *et al.*, 1968; Tikchonenko, 1969) or the Varian-produced model labeled as "cell-space total diffuse reflectance accessory for Cary 15" (Gratton, 1971). For the large and medium-sized bacteriophages the extrapolation-line slope is 3.9–4.0 (Tikchonenko *et al.*, 1966, 1974*a*; Tikchonenko, 1971; Gratton, 1971; Cancelieri *et al.*, 1974) and the UV phage absorption spectra obtained by extrapolation and using the integrating sphere are similar (Tikchonenko, 1969; Gratton, 1971). At the same time, Holzwarth *et al.* (1974) have calculated the light-scattering contribution to the T2-phage absorption spectra which deviates from the straight-line relationship in a certain spectral range, especially at 270–290 nm, leading to doubts about the validity of extrapolation at these wavelengths. However, the relationship is linear for other wavelengths, in particular, for 240–260 nm. It should be noted here that the integrating-sphere measurements of Sd and the T2 phage suspensions did not reveal any deviations from the linear dependence for the 270–290 nm spectral range (Rvachev *et al.*, 1968; Tikchonenko, 1969). Naturally, we cannot yet estimate the significance of these deviations and it is more prudent to wait for the promised detailed report [Dorman and Maestre, in preparation, cited in Holzwarth *et al.* (1974)].

For the phages with a high relative DNA content (40–50%) the protein contribution to the total absorption of the virus suspension at

260 nm is small, and, in addition, the protein absorption contribution may be easily accounted for since the empty capsids are stable and may be obtained as a purified preparation. The differential spectrum for the intraphage DNA is even easier to obtain by comparing the absorption spectra of the intact viral particles and the particles disintegrated with some mild procedure. In the latter case, the solution contains all the soluble virus components, including the internal proteins which are usually lost when preparing purified capsids. The optical properties of capsids in the isolated state and in the intact virions are almost the same due to the strong protein–protein interactions. In particular, the T2-phage fluorescence due to the protein component is completely preserved after applying mild disintegration techniques but changes significantly when drastic disintegration methods are used which lead to protein denaturation. This is the picture, for instance, observed in thermal disintegration of the T2 phage although the capsid preserves its normal morphology of the empty "ghost" (I. G. Charitonenkov, unpublished data).

UV absorption of the intraphage DNA is characterized, primarily, by incomplete hypochromism in the 250–290 nm range (Tikchonenko *et al.*, 1966; Inners and Bendet, 1969; Sinha and Misra, 1971; Gabrilovich *et al.*, 1968, 1970; Basu, 1971; Holzwarth *et al.*, 1974). Of special importance are the results reported by the last authors since they take into account the theoretically calculated light-scattering contribution. The deficit of hypochromism at 260 nm varies from 12 $\pm$ 1% for large phages of the T2 and Sd type to 6 $\pm$ 1% for the medium-sized phages of the λ and T7 type. These differences apparently reflect the individual characteristics of the DNP structure in various viruses.

In contrast to partially denaturated DNA preparations, which show approximately the same hyperchromism at all wavelengths, the intraphage DNA hypochromism grows with increasing wavelength. This is illustrated by the data for the Sd phage in Table 11.

This optical similarity may be explained by the properties of the intraphage DNA or by inadequate scattering correction. According to the first explanation the intraphage DNA resembles to a certain extent the preparations denatured in the presence of formaldehyde (Tikchonenko and Dobrov, 1969). The difference in absorption between the DNA preparations denatured in the presence and absence of HCHO is ascribed to the fact that formaldehyde reacts with the base amino groups yielding labile methylol derivatives distinguished by higher specific absorption in the 270–290 nm range (Haselkorn and Doty, 1961; Grossman *et al.*, 1961; Kislina and Tikchonenko, 1972).

TABLE 11

Sd-Phage–DNA Hyperchromism under Various Conditions

Conditions	Absorption[a]	at λ, nm		
	260	270	280	200
DNA in Sd virions	115	118	127	134
Sd DNA in solution after thermal denaturation	139	143	140	140
Sd DNA in solution after thermal denaturation and in the presence of formaldehyde	140	156	164	168

[a] Percentage of absorption of the native Sd DNA in solution at the same wavelength.

The higher absorption at longer wavelengths is characteristic also of the threadlike phages of the fd type (see above).

Of course, similarity of the specific absorption values for the intraphage DNA and for DNA after reaction with HCHO does not mean anything by itself but it may indicate a possible cause of the anomalous UV absorption spectrum of the intraphage DNA. Thus, it may be the involvement of a part of the base amino groups in the interaction with the protein component. The second possible cause may be a different conformation of the nucleoside residues in the regions of altered secondary structure, for instance, a possible change of the φC angle, leading in a limiting case, to the syn conformation (see above). Moreover, the observed anomalous absorption spectra of the phages may be caused not by one but by a number of factors.

Judging by the other possibility the long-wavelength dependence of hyperchromism value may be understood as a deviation of the λ *vs.* σ relationship from linearity in this spectral range. According to Maestre (personal communication) neither the extrapolating and integrating-sphere nor the fluorscat method can provide an adequate correction for the 270–300 nm region. He indicates that, even when corrected by fluorscat, the hyperchromism of DNA *in situ* has a wavelength behavior very similar to the anomalous index-of-refraction dispersion in the absorption bands which is, essentially, of the type $n^2 = A + B/(\lambda^2 - \lambda_0^2)$. This is suspiciously similar to the scattering-curve behavior. One must admit that inadequate light-scattering correction may influence, to a greater or lesser extent, the absorption of phages at various wavelengths. At the same time, it is difficult to ascribe the entire dependence to this cause. A similar long-wavelength dependence has been described for small, threadlike phages of the fd type which certainly have a quite different structure, size, and scattering ability (see above). Changes of the absorption value at 260 and 280 nm have been demonstrated in the experiments with HCHO, with the phage particles retaining their integrity and light-scattering ability (see below).

Disintegration of the phage particles and DNA release into solution are accompanied by restoration of normal optical properties, which is registered as the hypochromic effect. For various viruses its value at 260 nm has been shown to vary from 12 ± 1 to $6 \pm 1\%$ (Tikchonenko *et al.*, 1966; Tikchonenko, 1969; Inners and Bendet, 1969; Sinha and Misra, 1971; Gabrilovich *et al.*, 1968, 1970). The hypochromic effect increases in the long-wavelength spectral region according to higher specific absorption typical of the DNA structure *in situ*. The method of disintegration does not influence the hypochromic effect and a similar absorption decrease is observed when the phage is disintegrated by heating, osmotic shock, or sodium dodecyl sulfate and when the phage particles interact with isolated bacterial walls (Tikchonenko *et al.*, 1966; Velikodvorskaya *et al.*, 1968).

The absence of a hypochromic effect or its inadequate value is, as a rule, due to imperfect purification of the preparation, to a great amount of disintegrated particles, or to strong virus aggregation. Usually, in these cases tan α turns out to be considerably less than -4. Gratton (1971) reported that T2 phage preparations not subjected to ultrafiltration had tan α of -3.5 and that the true absorption spectra determined by extrapolation and with the integrating sphere did not agree in this case. After additional purification by ultrafiltration tan α returned to the normal value (-4) and the spectra determined with the two above methods became identical.

If we assume that the hypochromism deficit of 6–12% at 260 nm is due to disturbance of base stacking and the H-bond system, it follows that 18–30% of all intraphage DNA should be in this state in various phages.

Interestingly, the magnetic CD spectra of DNA in T2 and T5 phages (Maestre *et al.*, 1971) resemble to a certain extent the spectra of the preparations partially denatured by heating, but these observations have not yet been properly interpreted.

There is some reason to believe that the change of the MCD band intensity is due to a disturbance in the $n \rightarrow \pi^*$ transition and H bonding involving adenine residues in the DNA molecule. At the same time, the B2, B1u, and E1u bands in T2 and T5 phages behave quite differently, apparently reflecting a different mode of DNA packing in the heads of these phages. Certainly, these data cannot be interpreted in terms of trivial DNA denaturation *in vitro*. Disruption of the regular periodicity of the secondary structure has been observed also in the compact ψ DNA form *in vitro* (Jordan *et al.*, 1972) but it has not yet been reported in detail.

2.2.2b. Chemical Modification of the Intraphage DNA

The calculations discussed above are supported by the results of intraphage DNA modification using substances which "feel" the secondary structure of polynucleotides. Quantitative determinations of the intraphage DNA bases not protected by the interplanary or complementary interactions are made widely, using such compounds as O-methylhydroxylamine (OMHA), glyoxal, bisulfite, nitrous acid, and formaldehyde. Application of the water-soluble carbodiimide for this purpose [see in References Tikchonenko, (1969)] proved to be unsuccessful due to intensive protein modification and appearance of numerous DNA-protein cross-links. Both factors greatly hinder the quantitative determination of the modified base percentages.

The action of OMHA and bisulfite consists in nucleophilic attack at cytosine (Phillips *et al.*, 1966; Kochetkov and Budowsky, 1969) and, to a considerably lesser extent, at adenine. The attack is directed at the C^4 atom and the C^5–C^6 double bond of the pyrimidine ring perpendicular to the ring plane so that accessibility of cytosine in polynucleotides to OMHA depends primarily on the interplanary interactions. Therefore, under physiological conditions cytosine in the native double-helix DNA is not modified either by OMHA (Tikchonenko *et al.*, 1971) or by bisulfite (Skladneva *et al.*, 1973), whereas both compounds readily react with the intraphage DNA. When the Sd phage is modified by OMHA at 32°C the reaction takes 16–18 hours with 18 ± 2% of the intraphage cytosine reacting during this period. The remaining cytosine is, apparently, located in the regions with the perfect double helix and is, therefore, nonreactive. It should be noted that the reaction rate constant for the first reaction stage (formation of the intermediate product) for the intraphage DNA (0.2 hr^{-1}) proved to be close to the modification rate constant of free cytosine (0.18 hr^{-1}). In other words, cytosine located in the regions with altered secondary structure is not protected from nucleophilic attack by interplanary interactions and is similar to monomeric cytosine in this respect (Tikchonenko *et al.*, 1971).

Similar results were obtained with bisulfite, which modified 20% of the cytosine in the Sd phage (Skladneva *et al.*, 1973). At the same time, there is no simple relation between the reactivity of intraphage DNA bases toward OMHA, on one hand, and amount of DNA with altered secondary structure, on the other. A cause for such discrepancies observed with some Sd and λ phage mutants is a complicated mechanism of reaction *in situ* (Tikchonenko *et al.*, 1974*c*). In particular, OMHA can modify in the phage only cytosine residues interacting with nucleophilic amino acids (see below).

If about 20% of the intraphage cytosine is located in regions with disturbed base stacking, naturally the same amount of guanine not involved in the complementary interactions with the unstacked cytosine should be there too. The amount of unstacked guanine has been determined using the electrophilic agent glyoxal, which reacts with guanine producing the stable adduct which can be isolated from the modified DNA and determined quantitatively. According to Mazurenko *et al.* (1972), glyoxal modifies about 20% of guanine (more than 15 and less than 25%) in the Sd phage. More accurate determination of the reactive guanine has proved to be impossible due to procedural difficulties. Thus, according to the results of three independent chemical reactions, about 20% of the GC pairs in intraphage Sd DNA is reactive, i.e., it is not involved in ordered base stacking.

Generally, these results are confirmed by deamination, which also demonstrates the localization of adenine in the regions with disturbed base stacking. Unfortunately, the experimental conditions and reaction characteristics do not allow in this case the evaluation of the content of the unstacked bases in the intraphage DNA. The results discussed below improve the preliminary data on the kinetics of the deamination of intraphage DNA quoted in an earlier review (Tikchonenko, 1969).

The reactivity of base amino groups toward nitrosonium is not specific enough concerning the secondary structure, and both native and denaturated DNAs are deaminated (Litman, 1961; Schuster and Wilhelm, 1963; Kotaka and Baldwin, 1964). As the intraphage DNA includes regions with different secondary structure, the rate of intraphage DNA deamination on the whole is the mean value determined by the slower reaction of nitrosonium with the double-helix regions and by more rapid modification of the bases not protected by base stacking and complementary interactions (Tikchonenko *et al.*, 1966). As the phage-capsid permeability is not a limiting factor of the reaction, the deamination rates for the DNA double-helix regions *in vitro* and *in situ* should be nearly the same. Therefore, if the intraphage DNA contains preexisting free base amino groups the mean reaction rate *in situ* should be at the first stages higher than that for the native free DNA. In reality the situation is more complicated.

On the whole, the DNA of the Sd phage is deaminated with a

higher rate than that for the free DNA, thus proving the presence of regions with higher sensitivity to nitrosonium, but preexisting free amino groups are absent in the intraphage DNA (Kisseleva and Tikchonenko, 1972). As shown by the curves presented in Fig. 1, modification of intraphage adenine and cytosine not only lacks the rapid first stage, but it even starts after a 60- to 90-minute lag phase. Base deamination in the denatured DNA starts without any lag phase, and in native DNA a lag phase is observed only for cytosine deamination. Table 12 shows the rate constants for deamination of free and intraphage DNA.

A sharp (almost fivefold) increase is noticeable in the adenine deamination rate in the intraphage DNA. This is apparently due to the fact that among all the bases with NH_2 groups the sensitivity of adenine to nitrosonium is especially dependent on its conformation and involvement in various interactions. For instance, if the free adenine deamination rate (as a base) is taken to be unity, the adenine reaction rates in deoxyadenosine and native and denaturated calf thymus DNA are, respectively, 6.6, 0.44, and 0.63 (Shapiro and Pohl, 1968; Shapiro *et al.*, 1970). The higher deamination rates of cytosine and adenine in the intraphage DNA in conjunction with the presence of the lag phase indicate that, in addition to the double-helix regions, the intraphage DNA indeed has regions without ordered base stacking but that the amino

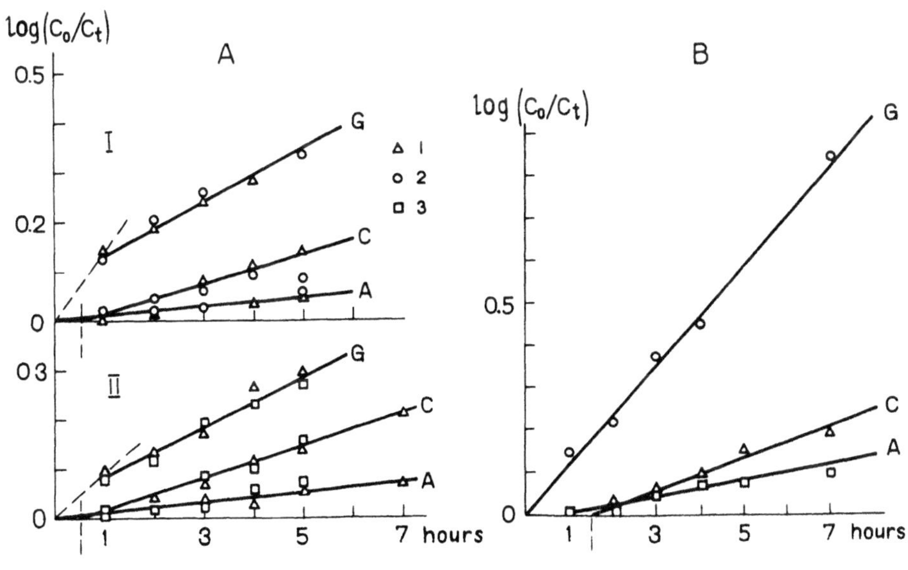

Fig. 1. Kinetics of base deamination in (A) free and (B) intraphage DNA; (AII and B) deamination in the presence of 0.17 M $Mg(CH_3COO)_2$, (AI) without Mg^{2+} 1 M $NaNO_2$, *p*H 4.15 ± 0.05, 37°C.

TABLE 12

Deamination Rate Constants[a]

DNA state	$K \times 10^3$ hr^{-1} A	C	G K_1	G K_2
Free Sd phage DNA	0.26 ± 0.02	0.89 ± 0.06	4.6 ± 0.35	2.0 ± 0.33
Sd intraphage DNA	1.2 ± 0.33	1.4 ± 0.32	4.7 ± 0.73	

[a] Incubation in 1 M $NaNO_2$ + 0.17 M Mg acetate, *p*H 4.2 ± 0.05, 37°C. K_1 and K_2 are the rate constants for the rapid and slow stages of guanine deamination, respectively.

group of these bases is, nevertheless, not free. Its blocking, apparently, is realized via a noncomplementary mechanism due to the interaction with protein postulated above. It should be stressed that the presence of a lag phase indicates certain order and cooperativity of such interactions, even though the base amino groups have weaker protection from the nitrosonium attack in this structure than in the double-helix DNA, as shown by the higher deamination rates.

The peculiar behavior of the C^2–NH_2 group of guanine during deamination of the free and intraphage DNA is determined by the extreme sensitivity of the third H bond in the GC pair to the acidic *p*H required for deamination [for details see the original publication (Kisseleva and Tikchonenko, 1972)].

2.2.3. Interaction between DNA and Protein in Phage Nucleoprotein

The above results concerning the special secondary structure of a part of the intraphage DNA imply that the interaction between DNA and protein is the factor determining the appearance of such a conformation. Apparently, this interaction cannot be limited to the conventional salt bonds between the negatively charged DNA phosphates and the positively charged protein amino acids which have been proved to exist in tissue nucleoproteins and in artificial complexes of DNA with alkaline proteins. The existence of these bonds may be confirmed by means of infrared spectrophotometry of DNA preparations at various humidities (Tsuboi, 1969). It has turned out that the vibration frequencies of the phosphate groups of free and intraphage DNA in the dry state are the same, and equal to 1245 cm^{-1} (Shie *et al.*, 1970,

1972*a*), indicating the absence in the phage of a noticeable amount of salt bonds between phosphates and protein amino groups (or imino groups). The above discussed optical data testify to the direct interactions between the bases and protein amino groups. This possibility is confirmed, primarily, by the results of chemical modifications, and in particular, by means of OMHA.

We have already discussed briefly the mechanism of the nucleophilic action of OMHA on the cytosine residues in the DNA molecule. This reaction has been sufficiently studied for monomers and polynucleotides (Kochetkov and Budowsky, 1969; Phillips *et al.*, 1966), but the reaction of OMHA with DNA in nucleoprotein complexes in the intact phage particles differs significantly from the modification of the free DNA (Tikchonenko *et al.*, 1971, 1974*a*). Figure 2 shows the scheme of the reaction between OMHA and cytosine, allowing for the peculiarities of DNA modification *in vitro* and *in situ*. The reaction *in vitro* proceeds in two directions: via the direct nucleophilic attack at the C^4 atom, giving rise to *N*-4-methoxycytidine (IV), and via the attack at the C^5–C^6 double bond, giving rise at first to the intermediate 6-methoxyamino 5,6-dihydrocytidine (II), which is further transformed into the second final product, *N*-4-methoxy-6-methoxyamino-5,6-dihydrocytidine (III), owing to addition of one more OMHA residue to the C^4 atom. The product IV is not found at all in modification of the phage Sd and the only final product identified is the product III, which

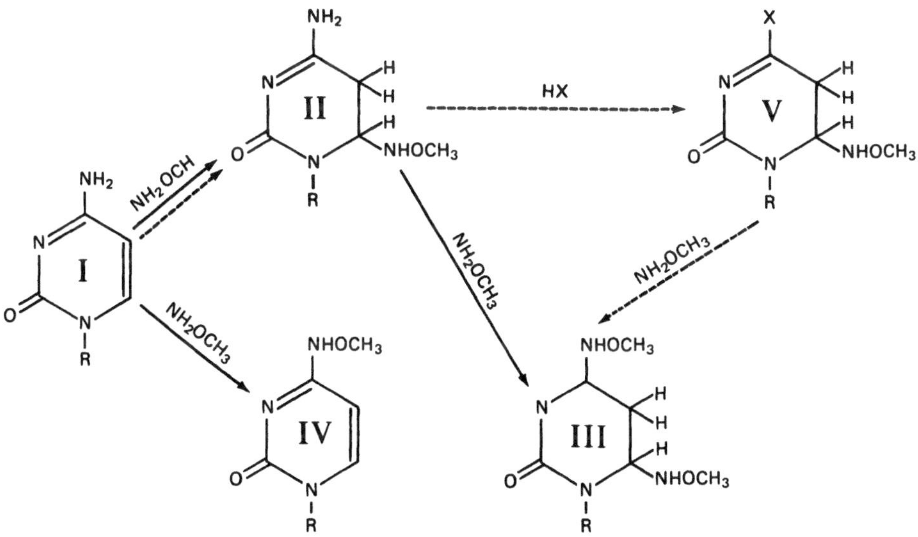

Fig. 2. The reaction between OMHA and cytosine. The solid arrows indicate reaction pathways *in vitro;* the dashed arrows, *in situ;* R, sugar residue; and XH, nucleophilic grouping of phage protein.

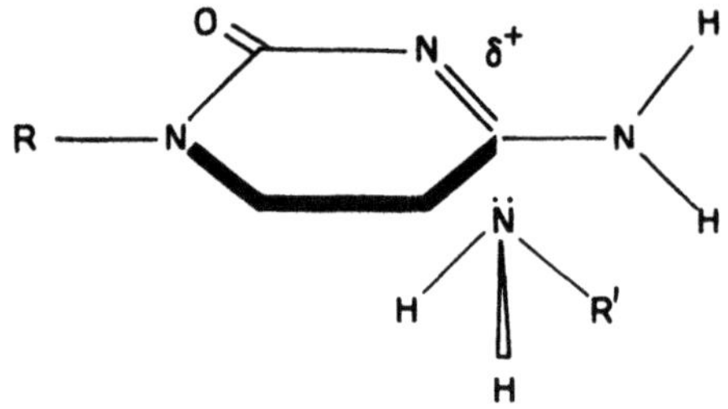

Fig. 3. Schematic structure of the cytosine–amino acid complex in intraphage DNA.

is formed via the additional intermediate product V, which is a cross-link between DNA and protein. OMHA modification of the T2 phage reveals even more marked deviations from the conventional reaction scheme (Khromov *et al.*, 1973). As 5-hydroxymethylcytosine has a hydroxymethyl group in the 5 position, making an attack at the C^5–C^6 double bond extremely difficult, the reaction of this anomalous base *in vitro* yields only a type IV product. Despite this, no traceable amounts of the product IV are formed in the T2 phage and the final modification product is a type III compound, formed, similar to the Sd phage, via the intermediate compound V.

A detailed analysis of the DNA modification mechanism in the phage nucleoprotein has shown that the C^4 cytosine atom is resistant to direct nucleophilic attack since all the reactive cytosine in the phage is complexed with amino acid, as illustrated in Fig. 3. Therefore, OMHA in the phage reacts not with cytosine *per se*, but with a cytosine–amino acid complex (cytosine in ordered base stacking is not reactive at all). Owing to the existence of this complex, the C^4 atom of the cytosine ring is resistant to direct nucleophilic attack, explaining the absence of product IV in the intraphage DNA. However, this atom is activated after saturation of the C^5–C^6 double bond, thus becoming sensitive to nucleophilic attack from the amino group of the nucleophilic amino acid protein residue. It should be noted that this occurs in the presence of OMHA, which is a much stronger nucleophilic agent and which finally displaces the amino acid from the product V, eliminating the cross-link between the protein and DNA. This is explained by the fact that in the starting DNP the amino acid interacted with protein and the cross-link appeared at the site of this preexisting interaction. Later this concept of transformation of the noncovalent interactions into the covalent cross-links was applied to study the DNA–protein interactions in the phages under UV irradiation (Simukova and Budowsky, 1974) and when modified by bisulfite (Skladneva *et al.*, 1973).

It should be noted that this mechanism of the DNA–protein interaction may explain also the resistance of the NH_2 base groups (and not only of cytosine!) to the above electrophilic agents (Tikchonenko and Dobrov, 1969; Tikchonenko *et al.*, 1974*a,b*; Kisseleva and Tikchonenko, 1972). This resistance is based both on the lowered reactivity due to electrons being drawn toward the ring, with the resultant redistribution of electron density, and on the steric factors determining the accessibility of a given functional group to the reactant.

The reversible cross-link appearing between DNA and protein during OMHA modification results in reversible phage inactivation, providing a fine analytic instrument to study the DNA–protein interactions in viruses (Andronikova *et al.*, 1974; Tikchonenko *et al.*, 1974*c*).

The base-stacking disturbance resulting from the interaction between nucleic acid and proteins is apparently not limited to viral nucleoproteins, but is a general phenomenon. By means of proton magnetic resonance spectroscopy Raszka and Mandel (1971) obtained convincing evidence that the interaction between polyadenylic acid and the aromatic amino acids leads to disruption of base stacking.

The protein participation in the interaction with the DNA bases opens a way for influencing the conformation of the intraviral nucleic acids. Indeed, if the protein NH_2 groups are responsible for disruption of base stacking in the phage DNP, the agents specifically acting on these functional groups could be capable, to a certain extent, of contributing toward restoration of the normal base stacking in DNA *in situ*.

Such conformational transition (see Table 13) resulting in restora-

TABLE 13

The Optical Properties of the Intraphage DNA under Various Conditions

Preparation	$\epsilon(P)$[a] Sd	$\epsilon(P)$[a] T2	$[\alpha]_{268}$ Sd	$[\alpha]_{268}$ T2[b]
Native DNA	6400	6600	+2300	+1800
Intact phage	7200	7400	+500	−1100
Phage after 25-min incubation with 4.5% HCHO at 40°C	6400	—	+2100	+1700
Phage after 25-min incubation with 4.5% HCHO at 40°C followed by dialysis	7100	—	+550	+400

[a] The values of A_t used for calculation $\epsilon(P)$ were determined both by extrapolation and in the integrating sphere (Tikchonenko *et al.*, 1974*a*).

[b] The T2 phage reveals incomplete restoration of the initial value of $[\alpha]_{268}$ after dialysis, apparently, due to a partial disintegration of the phage.

tion of normal optical properties of the DNA may indeed be carried out directly in the viral particles (Tikchonenko *et al.*, 1974*a,b*). This effect is realized by means of phage incubation which formaldehyde or glyoxal (3–4% solutions, 37–42°C, 20–60 min). The most interesting feature of this process is its reversibility; removal of formaldehyde from the medium results in restoration of the specific DNA conformation, with its hypochromism deficit and the small amplitude of the positive rotation peak.

The reversibility of the conformational alterations of the DNA *in situ* under the action of formaldehyde indicates that in the phage nucleoprotein it is protein that is permanently responsible for alterations of the secondary structure and optical properties of DNA. However, the interaction between the protein and the DNA is irreversibly disrupted, with strong disturbances of the DNA packing in the phage head and, possibly, with denaturing capsid alterations, so that the protein loses its capability to produce appropriate changes of the DNA conformation. As mentioned above, the thermal stability of the Sd phage capsid increases in 0.0015 M $NaNO_3$, and in this solvent the intraphage DNA may be melted without disintegrating the viral particles. The hyperchromic effect caused by heating disappears almost completely upon cooling, but the ORD curves retain the form typical of partially denaturated DNA (Dobrov *et al.*, 1972*a*; Andriashvili *et al.*, 1972).

The above results of the formaldehyde and glyoxal experiments demonstrate that it is the interaction with protein rather than the compact structure of the condensed DNA with its hypothetical anomalous light scattering that is responsible for the hypochromism deficit and for the amplitude decrease of the CD band or of the positive rotation peak. Considering the functional groups of the proteins responsible for these perturbations of the intraphage DNA structure, the experiments with the electrophilic agents may be, by themselves, treated in two ways: On one hand, it may be suggested that the protein NH_2 groups reacting under these conditions with formaldehyde or glyoxal are directly involved in such interactions, on the other hand, the interaction with DNA may involve some quite different protein functional groups (e.g., OH groups or the aromatic amino acid rings), and alterations of the optical properties of the intraphage DNA may be associated with the change in capsid structure due to formylation ol those amino groups which not interact directly with DNA. However, the above OMHA experiments, particularly the formation of the DNA–protein cross-links, tend to support the first approach.

The divalent cations, particularly Mg^{2+}, are an important factor for stabilizing the intraphage DNA–protein interactions. The above conformational transition may be effected only in the absence of Mg^{2+} (Tikchonenko *et al.*, 1974*b*). Incubation of the phages with either formaldehyde or glyoxal in the presence of 0.005–0.1 M Mg^{2+} does not restore *in situ* the optical properties typical of the native DNA in solution. In this case there occurs the normal chemical modification of the intraphage DNA accompanied by its characteristic optical effects. As should be expected in the presence of Mg^{2+}, the electrophilic agents react with the highest rate with the intraphage DNA regions having altered secondary structure and kept in contact with protein by divalent cations. The specific function of the Mg^{2+} ions in the phage DNP is to preserve the DNA–protein interactions. A similar picture is found in the tissue DNP [see references in Tikchonenko *et al.* (1974*b*)]. The Mg^{2+} ions, similar to some other divalent cations, are bonded to DNA as easily (Chang and Carr, 1968) as to proteins, the latter case apparently being due to the carboxyl groups (Malik and Agarwal, 1967). As a result, bridges may be formed in DNP between DNA and protein, apparently determining some stabilizing effect of the divalent cations on the structure and properties of chromosomes (Solari, 1965; Kirby, 1957).

2.2.4. The Biological Aspects of the Problem

Undoubtedly, the secondary structure of the DNA *in situ* directly influences the biological properties of phage particles. Evidently, inactivation and mutagenesis, physical stability of virions, and DNA injection, as well as some other properties of the viral particles, are, to a varying degree, associated with the structure of intraviral DNA and its interaction with protein. Mild chemical mutagenic agents such as OMHA, which are extremely sensitive to the secondary structure of the DNA and do not react at all with the double-helix DNA *in vitro* under physiological conditions, produce pronounced effects when acting on the intraphage DNA (Tikchonenko *et al.*, 1971, 1974*c*; Andronikova *et al.*, 1974; Khromov *et al.*, 1973). The differences in the mutagenic response of the free and intraphage DNA are undoubtedly also determined by the quantitative alterations of the base composition found after deamination *in situ* as compared to the experiments *in vitro* (Kisseleva and Tikchonenko, 1972).

An interesting approach to analysis of the viral DNP structure may be provided by the formation of a covalent bond between the DNA and the protein at the site of their preexisting interaction (see above) by means of OMHA or bisulfite. The cross-links between the DNA and protein appearing in the process naturally lead to virus inactivation due to the blocking of the DNA-injection mechanism (Andronikova *et al.*, 1974; Tikchonenko *et al.*, 1974*c*), which is reversible when OMHA is used. Biologically and genetically of great importance are the facts that different amino acids interact with cytosine *in situ* (Skladneva *et al.*, 1973) and that the reaction with OMHA involves not cytosine *per se* but its complex with an amino acid. As various amino acids in such complexes differ in their nucleophilicity, the rate of the intermediate product V formation and its further transformation into the final reaction product should depend significantly on the type of interacting amino acid. The cytosine complexes with amino acids devoid of nucleophilic properties will apparently turn out to be nonreactive or, more exactly, the reaction will terminate at the stage of the unstable product II (see Fig. 2). This should be followed by the development of additional structural microheterogeneity of the local DNP properties in the phage that would cause it to appear as though in the intraphage DNA there were several types of cytosine which differed in the reactivity.

Shown in Fig. 4 is the relationship between the time of incubation of the Sd phage with OMHA and the accumulation of reversible cross-

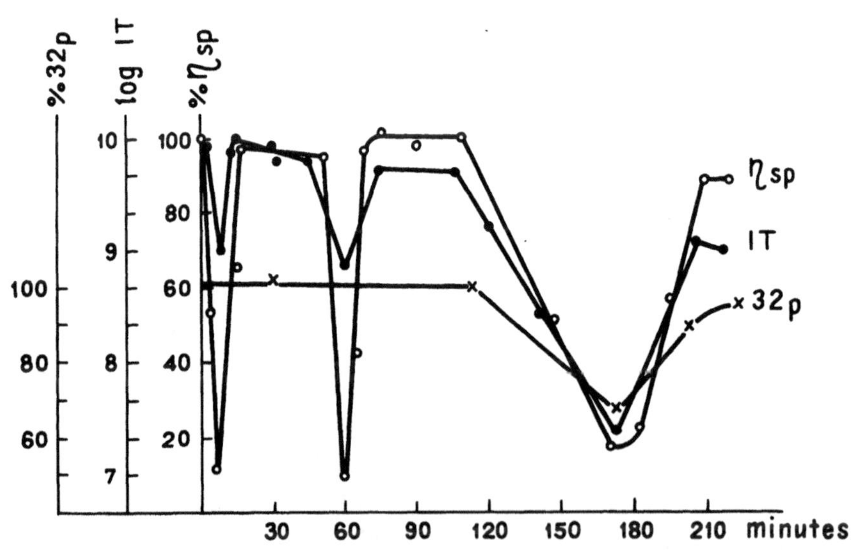

Fig. 4. Relationship between accumulation of reversible DNA–protein cross-links in the Sd phage and the time of incubation with OMHA. η_{sp}, viscosity of the preparations of phages disintegrated by heating; ^{32}P, injection of the ^{32}P-DNA from the phage particles into bacteria; IT, infectious titers of phages during incubation with OMHA. 1 M OMHA, *p*H 5.2, 32°C.

links between DNA and protein, which is determined using three independent techniques: measurements of viscosity of DNA released from disintegrated viral particles (Tikchonenko *et al.*, 1973), determination of injection of the phage ^{32}P-DNA into bacterial cells (Andronikova *et al.*, 1974), and determination of reversible phage inactivation (Tikchonenko *et al.*, 1974*c*). All three reversible effects (DNA-viscosity drop, DNA-injection blocking, and phage inactivation) are caused by the same event—reversible cross-linking of DNA with protein.

The curves presented show a number of successive periods of reversible inactivation of the phage Sd, indicating repeating, discrete, separate cycles of cross-linking and unlinking DNA and protein. A similar, though not identical, pattern is observed for the T7 phage (Andronikova *et al.*, 1974), whereas only one period of reversible inactivation has been observed for the λ and SW phages. However, reversible inactivation of phages with OMHA is extremely sensitive to various mutation changes. Shown in Fig. 5 are the inactivation curves for various mutants of the λ phage and their wild-type revertants (Tikchonenko *et al.*, 1974*c*). Interestingly, in part of these mutants the structural capsid proteins are altered (Tsus6, Gsus9, etc), whereas in others (λcI60 and λcI90) mutation involves only the intracellular, virus-specific proteins. In other words, mutations not associated directly with the virion capsid may, nevertheless, influence the interaction between DNA and protein in viral DNP. Therefore, it may be suggested that the fine chemical structure of the viral particle is determined not only by the primary structure of the capsid proteins and nucleic acid but also by the characteristics of the virion assembly during morphogenesis which may influence the DNA–protein interaction. This suggestion is supported by the data obtained by Fry and Weiters (1969) on the significant differences in the physical and even antigenic properties between the wild-type phage and its λcI60 mutant.

The quantitative and qualitative differences between the OMHA reversible inactivation curves for various phages and their mutants indicate considerable species and strain differences in the cytosine interactions with various amino acids. In this connection we should note several reversible inactivation periods observed for Sd and T7 phages. Such complex-inactivation patterns cannot be explained by the fact that each period corresponds to a special sort of complex of cytosine with a given nucleophilic amino acid. If this were so we could indeed expect each complex to have its own inactivation rate dependent on the nucleophilicity of its amino acid partner. However, it is clear that summation of several concurrent reactions with different rates would result in their averaging, and the resulting curve would not have several

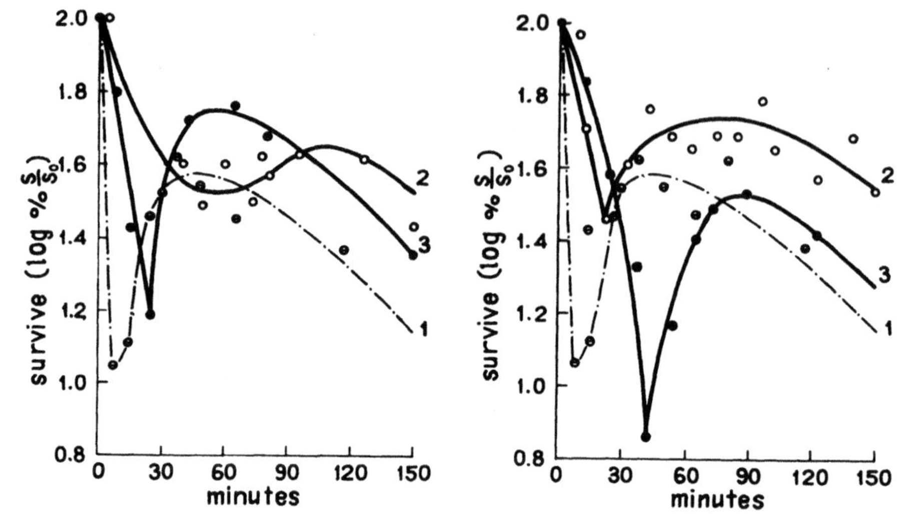

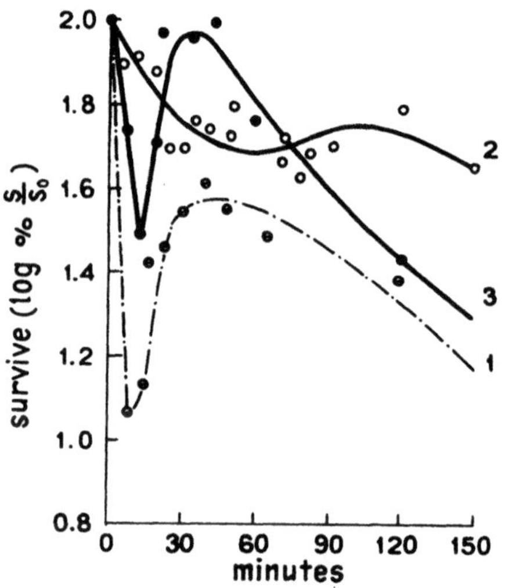

Fig. 5. Kinetics of the reversible inactivation of the λ phage mutants and their wild-type revertants at the early stages of the interaction with OMHA, 1 M OMHA, *p*H 5.2, 32°C. The wild-type revertants have been obtained either by spontaneous reversion or by incubation with OMHA (Andronicova *et al.*, 1974; Tikchonenko *et al.*, 1974*c*).

maxima and minima. Therefore, the nucleoprotein bonds in the phage particle should not be represented as the arithmetical mean of the interactions of the individual bases with individual amino acids. Most probably, such interactions should take place within more or less large blocks involving large DNA fragments and protein subunits, possibly even capsomers. In this connection, one can recall the bumps of the isometric viruses containing single-stranded RNA discussed in the Sect. 1 of this chapter. Such "group interactions" provide for specific recognition of DNA and protein [see von Hippel and McGee (1972) and Bourgeoi (1972)]. This possibility is indirectly confirmed by the results obtained by Sinha and Misra (1971), who have shown that the regions of altered structure in the T7 phage DNA consist of comparatively large fragments. This is supported also by the results of A. I. Zintchenko and N. P. Kisseleva (unpublished data) who have found that four different capsid proteins are cross-linked to the Sd phage DNA in the third period of reversible inactivation (see Fig. 4).

The postulated DNP blocks should include a considerable number of base–amino acid complexes, which, in turn, may interact between themselves, giving rise to elements of cooperativity. As a result, the majority of the cytidylic residues in a block will react with OMHA simultaneously and after considerable delay, which is necessary for the disruption of the cooperative interaction system in this block (Tikchonenko *et al.*, 1974*c*). In this case, the plots of the phage infective titer *vs.* the time of OMHA incubation reveal that the lag phase (the stationary region) is followed by the formation of DNA–protein cross-links all over the block, resulting in a distinct peak of reversible inactivation. This approach is in agreement with the above data on the lag phase during deamination of adenine and cytosine (Kisseleva and Tikchonenko, 1972). This may also be explained by DNP structural alterations in the block so that additional regions of DNA enter into contact with protein and nucleophilic amino acids (Tikchonenko *et al.*, 1973).

In conclusion we shall summarize the above experimental results in the form of 15 basic experimental observations, every one of which should be accounted for when constructing any structural model of the intraphage DNA.

1. The intraphage DNA is distinguished by a hypochromism deficit whose value at 260 nm varies from 6 to 12% for different phages studied, although these figures may be influenced by the scattering anomaly.

2. In contrast to the hyperchromic effect observed during the usual DNA denaturation in solution, the hypochromism deficit in the intraphage DNA grows with increasing wavelength. This dependence may also result to a lesser or greater degree from inadequate scattering correction.

3. The intraphage DNA is distinguished by anomalous ORD and CD spectra and, primarily, by a sharp decrease of the amplitude of the positive peak and by the red shift of the spectrum, even though there are significant differences in the optical activities of various phages.

4. In general, the ORD and CD spectra of intraphage DNA differ quantitatively and qualitatively from the spectra of both native and denaturated DNA.

5. Disintegration of virions and release of DNA are accompanied under favorable conditions by immediate renaturation.

6. Reversible "renaturation" of the normal optical properties of DNA may also be carried out *in situ* by incubation of phage suspension with formaldehyde or glyoxal in a medium which lacks Mg^{2+}.

7. A part of the intraphage DNA bases may be chemically modified by agents not reacting or poorly reacting with bases in the regular stacked array.

8. Free base amino groups are absent from intraphage DNA in spite of disruption of base stacking in part of polynucleotide chain.

9. The C^4 cytosine atom in the region of disturbed base stacking is insensitive to direct nucleophilic attack by OMHA.

10. The double bond of the pyrimidine ring in a part of the cytosine residues is sensitive to nucleophilic attack.

11. Cross-links between DNA and protein appear during phage modification with OMHA or bisulfite.

12. A lag phase exists during the intraphage DNA modification with OMHA and nitrous acid.

13. The sugar–phosphate skeleton of the whole molecule of intraphage DNA is undisturbed and the phosphate groups have normal hydration level.

14. The double-stranded parts of intraphage DNA show B or C configuration.

15. No salt bonds between the DNA phosphates and the capsid amino groups appear to exist.

The above observations may be accounted for in terms of the unified approach, suggesting that the DNA–protein interaction gives rise to a structural heterogeneity of the intraphage DNA, which has regions with altered secondary structure along with double-helix regions. The alteration is produced by a disturbance of base stacking due to the direct interaction between the bases and the protein amino acid residues. It has been suggested that the disturbance of base stacking with the sugar–phosphate skeleton being preserved is due to base rotation around the *N*-glycoside bond.

3. CONCLUDING REMARKS

1. The intimate interactions between nucleic acid and protein in the viral particles are an important factor, largely determining the biological properties of virions, their inactivation, sensitivity to mutagenesis, etc. The strength of such interaction and its contribution to the general physical stability of the nucleocapsids vary greatly in different viruses and depend to a great extent on the size, chemical composition, and symmetry type of the protein capsid and on the protein subunit structure.

2. The interaction between protein and nucleic acid in the viral nucleoproteins is not limited to the salt bonds between the negatively charged phosphates and positively charged protein groups, but it also includes the direct interaction between the bases and various functional protein groups. This seems to be the principal factor responsible for the conformational alterations of the nucleic acid structure *in situ*.

3. The maximum degree of interaction between protein and nucleic acid may be realized in the rodlike or filamentous viruses with spiral capsid symmetry where the single nucleotide strand is buried in protein following the packing of the protein subunits. According to this, the RNA in the particles of the tobacco mosaic virus type does not have *in situ* any inherent secondary structure based on the interplanary and complementary interactions. Rigid fixation of the intraviral RNA structure and chemical inertness of its bases are determined by its interaction with protein. Yet, in other filamentous viruses of the fd phage type the interaction between protein and single-stranded DNA exerts only a small influence on the polynucleotide-chain spiralization.

4. In the isometric viruses the nucleoprotein bonds are confined to the region adjoining the capsid, and, therefore, the effect of the nucleic acid–protein interaction on the structure of RNA or DNA is less pronounced. In the virions of the turnip yellow mosaic virus type, the tertiary structure of the RNA on the whole follows the icosahedral capsid symmetry, but the percentages of RNA helical structure *in situ* and *in vitro* either are the same or differ only slightly. In the phages of the Sd or T2 type containing double-stranded DNA, the major part of the polynucleotide chain has the Watson–Crick double-helix structure (B or C form) but 18–30% of the molecule has an altered configuration. In these regions the bases are drawn out of usual base stacking due, most probably, to the interaction with the protein amino groups.

5. The interaction between the bases and the protein amino acid residues gives rise to structural microheterogeneity of the nucleic acid and contributes toward variation of the reactivities of the RNA and DNA functional groups. In some cases the chemical modification of viruses involves not the bases *per se* but their complexes with the respective amino acids *in situ*.

4. REFERENCES

Adler, K., Beyreuther, K., Fanning, E., Geissler, N., Gronenborn, B., Klemm, A., Muller-Hill, B., Pfahl, M., and Schmitz, A., 1972, How *lac* repressor binds to DNA, *Nature* (*Lond.*) **237**, 322.

Akimenko, N. M., Djakowa, E. B., Evdokimov, Y. M., Frisman, E. V., and Varshavsky, Y. M., 1973, Viscosimetric study on compact form of DNA in water-salt solutions containing polyethylene glycol, *FEBS* (*Fed. Eur. Biochem. Soc.*) *Lett.* **38,** 61.

Alberts, B. M., Frey, L., and Delius, H., 1972, Isolation and characterization of gene 5 protein of filamentous bacterial viruses, *J. Mol. Biol.* **68,** 139.

Allen, F. S., and van Holde, K. E., 1971, Dichroism of TMV in pulsed electric fields, *Biopolymers* **10,** 865.

Anderegg, J. W., Geil, P. H., Beeman, W. W., and Kaesberg, P., 1961, An X-ray scattering investigation of wild cucumber mosaic virus and a related protein, *Biophys. J.* **1,** 657.

Anderegg, J. W., Wright, U., and Kaesberg, P., 1963, An X-ray scattering study of bromegrass mosaic virus, *Biophys. J.* **3,** 175.

Andriashvili, I. A., Dobrov, E. N., and Tikchonenko, T. I., 1972, Influence of low ionic strength and solute on the Sd phage stability and melting of intraphage DNA, *Biokhimia* **37,** 1251.

Andronikova, M. L., Velikodvorskaya, G. A., Tchruni, F. N., and Tikchonenko, T. I., 1974, The biological effects of chemical modification of the intraphage DNA by O-methylhydroxylamine, *Mol. Biol.* (*Moscow*) **8,** 3.

Asbeck, F., Beyreuther, K., Köhler, H., von Wettstein, G., and Braunitzer, G., 1969, Virus proteins, IV: The constitution of the coat protein of the fd phage, *Hoppe-Seyler Z. Physiol. Chem.* **350,** 1047.

Bachrach, H. L., 1964, Foot and mouth disease virus: Structure and mechanism of degradation as deduced from absorbance-temperature relationships, *J. Mol. Biol.* **8,** 348.

Bachrach, H. L., 1965, Foot and mouth disease virus: Structural changes during reaction with cations and formaldehyde as deduced from absorbance measurements, *Virology* **25,** 532.

Bachrach, U., and Friedmann, A., 1967, Purification and some possible functions of internal proteins from coliphage T2, *Biochem. Biophys. Res. Commun.* **26,** 596.

Bancroft, J. B., 1970, The self-assembly of spherical plant viruses, *Adv. Virus Res.* **16,** 99.

Bancroft, J. B., Hiebert, E., Rees, M. W., and Markham, R., 1968*a*, Properties of cowpea chlorotic mottle virus, its protein and nucleic acid, *Virology* **34,** 224.

Bancroft, J. B., Wagner, G. W., and Bracker, C. E., 1968*b*, The self-assembly of a nucleic acid free pseudo-top component for a small spherical virus, *Virology* **36,** 146.

Bancroft, J. B., Hiebert, E., and Bracker, C. F., 1969, The effects of various polyanions on shell formation of some spherical viruses, *Virology* **39,** 924.

Basu, S., 1971, Binding and interaction of acridine orange with intraphage DNA, *Biochim. Biophys. Acta* **254,** 48.

Basu, S., and Das Gupta, N. N., 1967, Spectrophotometric investigation of DNA in the ultraviolet, *Biochim. Biophys. Acta* **145,** 391.

Bawden, F. C., and Kleczkowski, A., 1959*a*, Photoreactivation of nucleic acid from tobacco mosaic virus, *Nature* (*Lond.*) **183,** 503.

Bawden, F. C., and Kleczkowski, A., 1959*b*, Some properties of decomposition products of potato virus X, *Virology* **7,** 375.

Belych, R. A., and Krivissky, A. S., 1966, Mutagenic action of nitrous acid on the ϕX174 phage infectious DNA, *Proc. Moscow Sci. Assoc.* (*Russ.*) **22,** 26.

Belych, R. A., Krivissky, A. S., and Tchernik, T. P., 1968, Comparison of mutagenic action of UV light on ϕX174 phage and its infectious DNA, *Genetika* **4,** 62.

Bendet, I., 1963, Biophysical characterization of bacteriophage nucleic acid. *Adv. Virus Res.* **10,** 65.

Bendet, I., and Mayfield, J. E., 1967, Ultraviolet dichroism of fd bacteriophage, *Biophys. J.* **7,** 111.

Bendet, I., Goldstein, D. A., and Lauffer, M. A., 1960, Evidence for internal organization of nucleic acid in T2 bacteriophage, *Nature, (Lond.)* **187,** 781.

Bendich, A., and Rosenkranz, G., 1963, Some thoughts on the double-stranded model of DNA, *Progr. Nucleic Acid Res.* **1,** 219.

Bishop, W. H., Quiocho, F. A., and Richards, F. M., 1966, The removal and exchange of metal ions in cross-linked crystals of carboxypeptidase-A, *Biochemistry* **5,** 4077.

Bode, O., and Paul, H. L., 1955, Elektronmikroskopische Untersuchungen über Kartoffel-Viren. I. Vermessungen an Teilchen des Kartoffel-X-Virus, *Biochim. Biophys. Acta* **16,** 343.

Boedtker, H., 1968, Dependence of the sedimentation coefficient on molecular weight of RNA after reaction with formaldehyde, *J. Mol. Biol.* **35,** 61.

Boeye, A., 1959, Induction of a mutation in poliovirus by nitrous acid, *Virology* **9,** 691.

Bonhoeffer, F., and Schachman, H. K., 1960, Studies on the organization of nucleic acids within nucleoproteins, *Biochem. Biophys. Res. Commun.* **2,** 366.

Bosch, L., Bonnet-Smits, A., and van Duin, J., 1967, *In situ* breakage of turnip yellow mosic virus RNA and *in situ* aggregation of the fragments, *Virology* **31,** 453.

Bouley, I. P., and Hirth, L., 1968, Action de la formamide sur le virus de la mosaique jaune du navef: Obtention de capsides arificielles, *C. R. Helod. Sennces Acad. Sci. Ser. D Sci. Nat.* **266,** 430.

Bourgeoi, S., 1972, Gene transcription of reproduction tissue, *in* "Karolinska Symposia on Research Methods in Reproductive Endocrinology," 5th Symposium, p. 178.

Boy de la Tour, E., and Kellenberger, E., 1965, Aberrant forms of the T-even phage head, *Virology* **27,** 222.

Bradley, D. E., 1965, The morphology and physiology of bacteriophages as revealed by the electron microscope, *J. R. Microscop. Soc.* **84,** (3), 257–316.

Bradley, D. E., 1967, Ultrastructure of bacteriophages and bacteriocins, *Bacteriol. Rev.* **31,** 230.

Brandes, J., and Bercks, R., 1965, Gross morphology and serology as a basis for classification of elongated plant viruses, *Adv. Virus Res.* **11,** 1.

Brown, G. D., and Zubay, G., 1960, Physical properties of the soluble RNA of *Escherichia coli.*, *J. Mol. Biol.* **2,** 287.

Brunner, W. C., and Maestre, M. F., 1974, Circular dichroism of films of polynucleotides, *Biopolymers* **13,** 345.

Buckingham, R. H., and Danchin, A., 1973, Fluorescence of tryptophanyl-tRNA^{Trp} from *E. coli:* An interaction between the indole and tRNA and its dependence on tRNA conformation, *FEBS (Fed. Eur. Biochem. Soc.) Lett.* **30,** 236.

Budowsky, E. L., Sherban, T. P., Krivissky, A. S., and Sverdlov, E. D., 1972, The effect of mutagenic agents on phage US2 and its infectious RNA. IV. The effect of *o*-methylhydroxylamine, *Genetika* **8,** 10, 63–73.

Bush, C. A., and Scheraga, H. A., 1967, Optical rotatory dispersion and RNA base pairing in ribosomes and in tobacco mosaic virus, *Biochemistry* **6,** 3036.

Cancellieri, A., Frontali, C., and Gratton, E., 1974, Dispersion effect on turbidimetric size measurement, *Biopolymers* **13,** 735.

Cantor, C. R., and Tinoco, I., Jr., 1965, Absorption and optical rotatory dispersion of seven trinucleoside diphosphates, *J. Mol. Biol.* **13,** 65.

Cantor, C. R., Jaskunas, S. R., and Tinoco, I., Jr., 1966, Optical properties of ribonucleic acid predicted from oligomers, *J. Mol. Biol.* **20,** 39.

Carpenter, J. M., and Kleczkowski, A., 1969, The absence of photoreversible pyri-

midine dimers in the RNA of ultraviolet-irradiated tobacco mosaic virus, *Virology* **39,** 542.
Caspar, D. L. D., 1956, Radial density distribution in the tobacco mosaic virus particle, *Nature* (*Lond.*) **177,** 928.
Caspar, D. L. D., 1962, Physical Principals in the construction of regular viruses, *Cold Spring Harbor Symp. Quant. Biol.* **27,** 1.
Caspar, D. L. D., 1963, Assembly and stability of the tobacco mosaic virus particle, *Adv. Protein Chem.* **18,** 37.
Caspar, D. L. D., and Klug, A., 1963, "Viruses, Nucleic Acid and Cancer," Williams & Wilkins, Baltimore.
Chang, C., Weiskopf, M., and Li, H. J., 1973, Conformational studies of nucleoprotein circular dichroism of deoxyribonucleic acid base pairs bound by polylysine, *Biochemistry* **12,** 3028.
Chang, K. Y., and Carr, C. W., 1968, The binding of calcium with deoxyribonucleic acid and deoxyribonucleic acid–protein complexes, *Biochim. Biophys. Acta* **157,** 127.
Cheng, P., 1968, Optical rotatory dispersion, tryptophan location, and base distribution in tobacco mosaic virus, *Biochemistry* **7,** 3367.
Chaproniere-Rickenberg, D. M., Mahler, H. R., and Fraser, D., 1964, The interaction of DNA and internal protein from coliphage T2, *Virology* **3,** 96.
Cohen, P., and Kidson, C., 1968, Conformational analysis of DNA–poly L-lysine complexes by optical rotatory dispersion, *J. Mol. Biol.* **35,** 241.
Cole, A., and Langley, R., 1963, Study of the radiosensitive structure of T2 bacteriophage using low energy electron beams, *Biophys. J.* **3,** 189.
Cram, L. S., and Deering, R. A., 1970, Ultraviolet inactivation dichroic ratio of oriented fd bacteriophage, *Biophys. J.* **10,** 413.
Crawford, L. V., 1966, A minute virus of mice, *Virology* **29,** 605.
Cummings, D. J., and Wanko, T., 1963, An electron microscopic study of T2 bacteriophage in thin sections, *J. Mol. Biol.* **7,** 658.
Cummings, D. J., Chapman, V. A., and De Long, S. S., 1965, An electron microscopic study of λ and λdg bacteriophage in thin sections, *J. Mol. Biol.* **14,** 418.
Damirdagh, I. S., and Shepherd, R. J., 1970, Some of the chemical properties of the tobacco etch virus and its protein and nucleic acid components, *Virology* **40,** 84.
Davidson, B., and Fasman, G., 1969, The double-stranded polyadenylic acid–poly-L-lysine complex. A conformational study and characterization, *Biochemistry* **8,** 4116.
Day, L. A., 1966, Protein conformation in fd bacteriophage as investigated by optical rotatory dispersion, *J. Mol. Biol.* **15,** 395.
Day, L. A., 1969, Conformations of single-stranded DNA and coat protein in fd bacteriophage as revealed by ultraviolet absorption spectroscopy, *J. Mol. Biol.* **39,** 265.
Day, L. A., 1973, Circular dichroism and ultraviolet absorption of a deoxyribonucleic acid-binding protein of filamentous bacteriophage, *Biochemistry* **12,** 5329.
Dembo, A., Dobrov, E. N., Lednev, V., Tikchonenko, T. I., and Feigin, L. A., 1965, About packing of DNA in the head of DD VII, T2 and Sd phages, *Biofizika* **10,** 404.
Dityatkin, S. Y., Danileytchenko, V. V., Zavilgelsky, G. V., and Ilyashenko, B. N., 1967, Comparison of UV lightsensitivity of Iφ7 and T7 phages and their infectious DNA, *Genetika* **11,** 87.
Dobrov, E. N., Andriashvili, I. A., and Tikchonenko, T. I., 1972*a*, The optical rotary study of Sd phage in low ionic strength solution, *Biokhimia* **37,** 1088.
Dobrov, E. N., Kust, S. V., and Tikchonenko, T. I., 1972*b*, The structure of single-stranded virus RNA *in situ*. A study of absorption spectra and optical rotatory

dispersion of tobacco mosaic virus and potato virus X preparations, *J. Gen. Virol.* **6**, 161.

Dobrov, E. N., Mazhul, L. A., Kust, S. V., and Tikchonenko, T. I., 1973, A study of the effect of ethylene glycol on some helical plant viruses, *Mol. Biol.* (*Moscow*) **7**, 254.

Dobrov, E. N., Lyaser, P. M., and Kust, S. V., 1974, Some optical properties of dolihos mosaic virus, *in* "Structure and Functions of Nucleic Acids and Nucleoproteins," Thesis of A. N. Belozersky Symposium Jan. 29–Feb. 11, 1974, Moscow State University, Moscow.

Dore, E., Frontali, C., and Gratton, E., 1972, Physico-chemical description of a condensed form of DNA, *Biopolymers* **11**, 443.

Dore, E., Frontali, C., and Notargiacomo, S., 1973, Electron microscopic observations of DNA condensates at low *p*H values, *J. Mol. Biol.* **78**, 391.

Dorman, B. P., and Maestre, M. F., 1973, Experimental differential light-scattering correction to the circular dichroism of bacteriophage T2, *Proc. Natl. Acad. Sci. USA* **70**, 255.

Doty, P., and Steiner, R. F., 1950, Light scattering and spectrophotometry of colloidal solutions, *J. Chem. Phys.* **18**, 1211–1220.

Doty, P., Boedtker, H., Fresco, J., Haselkorn, R., and Litt, M., 1959, Secondary structure in ribonucleic acids, *Proc. Natl. Acad. Sci. USA* **45**, 482.

Dunn, D., and Smith, J. D., 1958, *Abstr. 4th Interntl. Congr. Biochem. Vienna*, p. 72.

Dusenbery, D. A., and Uretz, R. B., 1972, The interaction of acridine dyes with the densely packed DNA of bacteriophage, *Biophys. J.* **12**, 1056.

Eiserling, F. A., and Dickson, R. C., 1972, Assembly of viruses, *Annu. Rev. Biochem.* **41**, 467.

Eisinger, J., 1966, Information Exchange Group, N 7 (JEG-7).

Englander, S. W., and Epstein, H. T., 1957, Optical methods for measuring nucleoprotein and nucleic acid concentration, *Arch. Biochem. Biophys.* **68**, 144.

Evdokimov, Y. M., Platonov, A. L., Tikchonenko, A. S., and Varshavsky, Y. M., 1972, A compact form of double-stranded DNA in solution, *FEBS* (*Fed. Eur. Biochem. Soc.*) *Lett.* **23**, 180.

Evdokimov, Y. M., Akimenko, N. M., Gluchova, N. E., Tikchonenko, A. S., and Varshavsky, Y. M., 1973, Formation of the compact form of double-stranded DNA in solution in the presence of polyethylene glycol, *Mol. Biol.* (*Moscow*) **7**, 151.

Evdokimov, Y. M., Akimenko, N. M., Gluchova, N. E., and Varshavsky, Y. M., 1974, DNA compact from in solution I., *Mol. Biol.* (*Moscow*) **8**, 396.

Falk, M., Hartman, K. A., and Lord, R. C., 1962, Hydration of deoxyribonucleic acid, *J. Am. Chem. Soc.* **84**, 3843.

Fasman, G. D., Schaffhausen, B., Goldsmith, L., and Adler, A., 1970, Conformational changes associated with f-1 histone deoxyribonucleic acid complexes. *Biochemistry* **9**, 2814.

Finch, J. G., 1965, Preliminary X-ray diffraction studies on tobacco rattle and barley stripe mosaic viruses, *J. Mol. Biol.* **12**, 612.

Finch, J. T., and Klug, A., 1966, Arrangement of protein subunits and the distribution of nucleic acid in turnip yellow mosaic virus, *J. Mol. Biol.* **15**, 344.

Finch, J. T., and Klug, A., 1967, Structure of broad bean mottle virus. I. Analysis of electron micrographs and comparison with turnip yellow mosaic virus and its top component, *J. Mol. Biol.* **24**, 289.

Finch, J. T., Klug, A., and van Regenmortel, M. H. V., 1967*a*, The structure of cucumber mosaic virus, *J. Mol. Biol.* **24,** 303.
Finch, J. T., Leberman, R., and Berger, J. E., 1967*b*, Structure of broad bean mottle virus. II. X-ray diffraction studies, *J. Mol. Biol.* **27,** 17.
Fischbach, F. A., Harrison, P. M., and Anderegg, J. W., 1965, An X-ray scattering study of the bacterial virus R17, *J. Mol. Biol.* **13,** 638.
Fraenkel-Conrat, H., 1954, Reaction of nucleic acid with formaldehyde, *Biochim. Biophys. Acta* **15,** 308.
Fraenkel-Conrat, H., 1969, "The Chemistry and Biology of Viruses," Academic Press, New York.
Fraenkel-Conrat, H., and Colloms, M., 1967, Reactivity of tobacco mosaic virus and its protein toward acetic anhydride, *Biochemistry* **6,** 2740.
Fraenkel-Conrat, H., and Singer, B., 1964, Reconstitution of tobacco mosaic virus. IV. Inhibition of enzymes and other proteins, and use of polynucleotides, *Virology* **23,** 354.
Frank, H., and Day, L. A., 1970, Electron microscopic observations on fd bacteriophage, its alkali denaturation products and its DNA, *Virology* **2,** 144.
Franklin, R., 1955*a*, Structure of tobacco mosaic virus, *Nature* (*Lond.*) **175,** 379.
Franklin, R., 1955*b*, Structural resemblance between Schramm's repolymerised A-protein and tobacco mosaic virus, *Biochim. Biophys. Acta* **18,** 313.
Franklin, R., 1956*a*, Location of the ribonucleic acid in the TMV particle, *Nature* (*Lond.*) **177,** 928.
Franklin, R., 1956*b*, X-ray diffraction studies of cucumber virus 4 and three strains of tobacco mosaic virus, *Biochim. Biophys. Acta* **19,** 203.
Franklin, R. and Klug, A., 1956, The nature of the helical groove on the tobacco mosaic virus particle, *Biochim. Biophys. Acta* **19,** 403.
Franklin, R., Klug, A., and Holmes, K. C., 1957, *in* "Nature of Viruses" (C.E.W. Wolstenholme and E.C.P. Miller, eds.), p. 39, Churchill, London.
Franklin, R., Caspar, D. L. D., and Klug, A., 1959, Problems and progress 1908–1958, *in* "Plant Pathology" (C. S. Holton, ed.), p. 444, University of Wisconsin Press, Madison, Wisc.
Fraser, R. D. B., 1952, Infra-red dichroism nucleoprotein tobacco mosaic virus, *Nature* (*Lond.*) **170,** 490.
Friedman, S., and Ts'o, P. P., 1971, Interaction of poly-L-tyrosine with nucleic acids. I. Formation of complexes, *Biochemistry* **10,** 3099.
Frisman, E. V., Vorobjev, V. I., Yanovskaya, N. K., and Shagina, L. V., 1963, The DLP study of molecular structure of ribonucleic acid, *Biokhimia* **28,** 137.
Fuller, W., 1961, Two-stranded helical configurations for ribonucleic acid, *J. Mol. Biol.* **3,** 175.
Fuller, W., Wilkins, M. H. F., Wilson, H. R., and Hamilton, L. D., 1965, The molecular configuration of deoxyribonucleic acid. IV. X-ray diffraction study of the A form, *J. Mol. Biol.* **12,** 60.
Furuse, K., and Watanabe, I., 1971, Effects of ultraviolet light (UV) irradiation on RNA phage in H_2O and in D_2O, *Virology* **46,** 171.
Gabbay, E. J., Sanford, K., and Baxter, C. S., 1972, Specific interaction of peptides with nucleic acids, *Biochemistry* **11,** 3429.
Gabbay, E. J., Sanford, K., Baxter, C. S., and Kapicak, L., 1973, Specific interaction of peptides with nucleic acids. Evidence for a "selective bookmark" recognition hypothesis, *Biochemistry* **12,** 4021.

Gabler, R., and Bendet, I., 1972, Comparison of the UV flow dichroism spectra of TMV and several of its mutants, *Biopolymers* **11,** 2393.

Gabrilovich, I. M., Polupanov, V. S., and Anisimova, N. I., 1968, Macromolecular structure of phage *Klebsiella* DNA, *Mol. Biol.* (*Moscow*) **2,** 155.

Gabrilovich, I. M., Romanovskaya, L. N., Zentchenko, S. A., and Resnikova, I. V., 1970, The interaction of the acridine dyes with DNA in solution and inside phage particles, *Mol. Biol.* (*Moscow*) **4,** 324.

Gellert, M., and Davies, D. R., 1964, Organization of DNA in bacteriophage T4, *J. Mol. Biol.* **8,** 341.

Gendon, Y. Z., 1966, Mutagenic and inactivating effect of hydroxylamine in treatment of infectious RNA and native polyomielitis virus, *Vopr. Virusol.* **6,** 724.

Ginoza, W., 1958, Kinetics of heat inactivation of ribonucleic acid of TMV, *Nature* (*Lond.*) **181,** 958.

Goddard, J., Streeter, D., Weber, C., and Gordon, M. P., 1966, Studies on the inactivation of tobacco mosaic virus by ultraviolet light, *Photochem. Photobiol.* **5,** 213–222.

Gomatos, P. J., Klug, R. M., and Tamm, I., 1964, Enzymic synthesis of RNA with reovirus RNA as template. I. Characteristics of the reaction catalyzed by the RNA polymerase from *Escherichia coli, J. Mol. Biol.* **9,** 193.

Gordon, D. J., 1972, Mie scattering by optically active particles, *Biochemistry* **11,** 413.

Gordon, D. J., and Holzwarth, G., 1971, Artifacts in the measured optical activity of membrane suspensions, *Arch. Biochem. Biophys.* **142,** 481.

Gorin, A. S., Spitkovsky, D. M., Tikchonenko, T. I., and Tseytlin, P. I., 1967, The secondary structure of DNA in phage particles, *Biochim. Biophys. Acta* **134,** 490.

Gottesfeld, J. M., Calvin, M., Cole, R. D., Idgaloff, D. M., Moses, V., and Vaughan, W., 1972, An investigation of specific interactions of deoxyribonucleic acid and lysine-rich (F1) histone preparations, *Biochemistry* **11,** 1422.

Gratton, E., 1971, Method for the automatic correction of scattering in absorption spectra by using the integrating sphere, *Biopolymers* **10,** 2629.

Green, G., and Mahler, H. R., 1970, Comparative study of polyribonucleotides in aqueous and glycol solutions, *Biochemistry* **9,** 368.

Greve, J., and Blok, J., 1973, Transient birefringence of T-even bacteriophages. I. T4B in the absence of tryptophan and fiberless T4 particles, *Biopolymers* **12,** 2607–2622.

Griffith, J. T., and Kornberg, A., 1972, *in* "Membrane Research" (C. F. Fox, ed.), p. 281, Academic Press, New York.

Grossman, L., Levine, S., and Allison, W. S., 1961, The reaction of formaldehyde with nucleotides and T2 bacteriophage DNA, *J. Mol. Biol.* **3,** 47.

Harrison, B. D., Finch, J. T., Gibbs, A. J., Hollings, M., Shepherd, R. J., Valenta, V., and Wetter, C., 1971, Sixteen groups of plant viruses, *Virology* **45,** 356.

Hart, R. G., 1955, Electron-microscopic evidence for the localization of ribonucleic acid in the particles of TMV, *Proc. Natl. Acad. Sci. USA* **1,** 261.

Haselkorn, R., 1962, Studies on infectious RNA from turnip yellow mosaic virus, *J. Mol. Biol.* **4,** 357.

Haselkorn, R., and Doty, P., 1961, The reaction of formaldehyde with polynucleotides, *J. Biol. Chem.* **236,** 2738.

Haynes, M., Garrett, R. A., and Gratzer, W. B., 1970, Structure of nucleic acid–poly base complexes, *Biochemistry* **9,** 4410.

Heisenberg, M., 1966, Formation of defective bacteriophage particles by fr amber mutants, *J. Mol. Biol.* **17,** 136.

Hélène, C., 1971, Role of aromatic amino-acid residues in the binding of enzymes and proteins to nucleic acids, *Nat. New Biol.* **234,** 120.

Hélène, C., and Dimicoli, J.-L., 1972, Interaction of oligopeptides containing aromatic amino acids with nucleic acids. Fluorescence and proton magnetic resonance studies, *FEBS (Fed. Eur. Biochem. Soc.) Lett.* **26,** 6.

Hélène, C., Dimicoli, J.-L., and Brun, F., 1971*a*, Binding of tryptamine and 5-hydroxytryptamine (serotonin) to nucleic acids. Fluorescence and proton magnetic resonance studies, *Biochemistry* **10,** 3802.

Hélène, C., Montenay-Garestier, A., and Dimicoli, J.-L., 1971*b*, Interactions of tyrosine and tyramine with nucleic acids and their components. Fluorescence, nuclear magnetic resonance and circular dichroism studies, *Biochim. Biophys. Acta* **254,** 349.

Henkens, R. W., and Middlebrook, J. L., 1973, Optical and hydrodynamic studies of the structure of bacteriophage f2, *Biochemistry* **12,** 2910.

Hiebert, E., Bancroft, J. B., and Bracker, C. E., 1968, The assembly *in vitro* of some small spherical viruses, hybrid viruses, and other nucleoproteins, *Virology* **34,** 492.

Hill, J. H., and Shepherd, R. J., 1972, Molecular weights of plant virus coat proteins by polyacrylamide gel electrophoresis, *Virology* **47,** 817.

Hoffmann-Berling, V. H., Marvin, D. A., and Dürwald, H., 1963, Ein fädiger DNS-phage (fd) und ein sphärischer RNS-phage (fr), wirtsspezifisch für mämmliche Stämme von *E. coli.* 1. Präparation und chemische Eigenschaften von fd und fr, *Z. Naturf. Orsch* **18B,** 876.

Hohn, T., 1969, Role of RNA in the assembly process of bacteriophage fr, *J. Mol. Biol.* **43,** 191.

Hohn, T., and Hohn, B., 1970, Structure and assembly of simple RNA bacteriophages, *Adv. Virus Res.* **16,** 43.

Holmes, K. C., and Franklin, R. E., 1958, The radial density distribution in some strains of tobacco mosaic virus, *Virology* **6,** 328.

Holzwarth, G., Gordon, D. G., McGinness, J. E., Dorman, B. P., and Maestre, M. F., 1974, Mie scattering contributions to the optical density and circular dichroism of T2 bacteriophage, *Biochemistry* **13,** 126.

Hosszu, J. L., and Rahn, R. O., 1967, Thymine dimer formation in DNA between 25°C and 100°C, *Biochem. Biophys. Res. Commun.* **29,** 327.

Huang, C. W., and Gordon, M. P., 1972, Photoreactivation of tobacco mosaic virus and potato virus X ribonucleic acid inactivated by acetone-sensitized photoreactivation, *Photochem. Photobiol.* **15,** 493.

Huang, C. W., and Gordon, M. P., 1974, The formation of photoreversible cyclobutane-type pyrimidine dimers in ultraviolet-irradiated potato virus X, *Photochem. Photobiol.* **19,** 269.

Hurter, J., Gordon, M. P., Kirwan, J. P., and McLaren, A. D., 1974, *In vitro* photoreactivation of ultraviolet-inactivated ribonucleic acid from tobacco mosaic virus, *Photochem. Photobiol.* **19,** 185.

Huxley, H. E., and Zubay, G., 1961, Preferential staining of nucleic acid-containing structures for electron microscopy, *J. Biophys. Biochem. Cytol.* **11,** 273–296.

Ikehara, K., Obata, Y., Utyama, H., and Kurata, M., 1973, *Bull. Inst. Chem. Res. Kyoto Univ.* **51,** 140.

Inman, R. B., and Jordan, D. O., 1960, The UV-absorption of calf-thymus DNA, *Biochim. Biophys. Acta* **42,** 530.

Inners, D., and Bendet, I. J., 1969, Thermal stability of T2 DNA *in situ, Virology* **38,** 269.

Inoue, S., and Ando, T., 1970, Interaction of clupeine with deoxyribonucleic acid. II. Optical rotatory dispersion studies, *Biochemistry* **9,** 395.

Isenberg, H., Cotter, R. I., and Gratzer, W. B., 1971, Secondary structure and interaction of RNA and protein in a bacteriophage, *Biochim. Biophys. Acta* **232,** 184.

Jacobsen, J., and Wang, J. C., 1974, On the possibility of intercalation of aromatic amino acid residues into double-stranded DNA helix, *Biochim. Biophys. Acta* **335,** 49.

Jacobsen, M. F., and Baltimore, D., 1968, Morphogenesis of poliovirus. I. Association of the virus RNA with coat protein, *J. Mol. Biol.* **33,** 368.

Jonard, G., 1972, Ph.D. Thesis, University of Strasburg, France.

Jonard, G., and Hirth, L., 1966, Action de l'urée sur le virus de la mosaique jaune de navet: Formation de capsides artificielles, *Ct. R. Hebd. Seances Acad. Sci. Ser. D Sci. Nat.* **236,** 1909.

Jonard, G., Ralijoana, D., and Hirth, L., 1967, Action de l'urée sur la virus de la mosaique jaune du navet. Properties du RNA obtenu lors de la formation de capsides artificielles par M.U, *Ct. R. Hebd. Seances Acad. Sci. Ser. D. Sci. Nat.* **264,** 2694–2698.

Jonard, G., Witz, J., and Hirth, L., 1972, Formation of nucleoprotein complexes from dissociated turnip yellow mosaic virus RNA and capsids at low *p*H: Preliminary observations, *J. Mol. Biol.* **67,** 165.

Jordan, C. F., Lerman, L. S., and Venable, J. H., Jr., 1972, Structure and circular dichroism of DNA in concentrated polymer solutions, *Nat. New Biol.* **236,** 67.

Kaper, J. M., 1968, The small RNA viruses of plants, animals and bacteria. A. Physical properties, *in* "Molecular Basis of Virology" (H. Fraenkel-Conrat, ed.), p. 1, Academic Press, New York.

Kaper, J. M., 1969, Nucleic acid-protein interactions in turnip yellow mosaic virus, *Science (Wash., D.C.)* **166,** 248.

Kaper, J. M., 1971, Studies on the stabilizing forces of simple RNA viruses. I. Selective interference with protein–RNA interactions in turnip yellow mosaic virus, *J. Mol. Biol.* **56,** 259.

Kaper, J. M., 1972, RNA viruses: Replication and structure, *FEBS (Fed. Eur. Biochem. Soc.) Symp.* **27,** 19.

Kaper, J. M., 1973, Arrangement and identification of simple isometric viruses according to their dominating stabilizing interactions, *Virology* **55,** 299.

Kaper, J. M., and Geelen, J. L. M. C., 1971, Studies on the stabilizing forces of simple RNA viruses. II. Stability, dissociation and reassembly of cucumber mosaic virus, *J. Mol. Biol.* **56,** 277.

Kaper, J. M., and Halperin, J. E., 1965, Alkaline degradation of turnip yellow mosaic virus. II. *In situ* breakage of the ribonucleic acid, *Biochemistry* **4,** 2434.

Kaper, J. M., and Jenifer, F. G., 1965, Studies on the interaction of *p*-chloromercuribenzoate with turnip yellow mosaic virus. III. Involvement of the ribonucleic acid, *Arch. Biochem. Biophys.* **112,** 331.

Kaper, J. M., and Jenifer, F. G., 1967, Studies on the interaction of *p*-mercuribenzoate with turnip yellow mosaic virus. IV. Conformational change, exposure of buried prototropic groups, and *p*-H-induced degradation, *Biochemistry* **6,** 440.

Kaper, J. M., and Jenifer, F. G., 1968, Studies on the interaction of *p*-mercuribenzoate with turnip yellow mosaic virus. V. Induced ribonuclease sensitivity and degradation of the virion, *Virology* **5,** 71.

Kaper, J. M., Diener, T. O., and Scott, H. A., 1965, Some physical and chemical properties of cucumber mosaic virus (strain Y) and of its isolated ribonucleic acid, *Virology* **27**, 54.

Kassanis, B., and Kleczkowski, A., 1965, Inactivation of a strain of tobacco necrosis virus and of the RNA isolated from it by UV radiation of different wave-lengths, *Photochem. Photobiol.* **4**, 209.

Katz, L., and Rich, A., 1966, X-ray diffraction study of large phages, *Abstr. Biophys. Soc. USA 10th Annu. Meet. Boston,* p. 58.

Kausche, G. A., and Hahn, F., 1948, Über die stöchiometrische Farbstoffverbindungen des Tabakmosaikvirusproteins, *Z. Naturforsch.* **3B**, 437–441.

Khromov, I. S., Ogarova, N. L., and Tikchonenko, T. I., 1973, *in* "Molecular Biology of Viruses," p. 80, Academy of Medical Sciences & Institute of Virology, Moscow.

Kilkson, R., 1957, Cylindrically averaged electron density distribution in cucumber virus number four, *Arch. Biochem. Biophys.* **67**, 53.

Kilkson, R., and Maestre, M. F., 1962, Structure of T-2 bacteriophage, *Nature* (*Lond.*) **195**, 494.

Kirby, K. S., 1957, A new method for the isolation of deoxyribonucleic acids: Evidence on the nature of bonds between deoxyribonucleic acid and protein, *Biochem. J.* **66**, 495.

Kislina, O. S., and Tikchonenko, T. I., 1972, Interaction of formaldehyde with intraphage DNA, *Biokhimia* **37**, 372.

Kisseleva, N. P., and Tikchonenko, T. I., 1972, Kinetics of diamination of Sd phage DNA *in situ* and in solution by nitrous acid, *Biokhimia* **37**, 562.

Kleczkowski, A., and Govier, D. A., 1969, Action spectrum for inactivation of the infectivity of potato virus X by U.V. radiation, *Photochem. Photobiol.* **10**, 53.

Kleczkowski, A., and McLaren, A. D., 1967, Inactivation of infectivity of RNA of TMV during ultraviolet-irradiation of the whole virus at two wavelengths, *J. Gen. Virol.* **1**, 441.

Klimenko, S. M., Tikchonenko, T. I., and Andreev, V. M., 1967, Packing of DNA in the head of bacteriophage T2, *J. Mol. Biol.* **23**, 523.

Klug, A., and Caspar, D. L. D., 1960, The structure of small viruses, *Adv. Virus Res.* **7**, 225.

Klug, A., and Finch, J. T., 1960, The symmetries of the protein and nucleic acid in turnip yellow mosaic virus: X-ray diffraction studies, *J. Mol. Biol.* **2**, 201.

Klug, A., Holmes, K. C., and Finch, J. T., 1961, X-ray diffraction studies on ribosomes from various sources, *J. Mol. Biol.* **3**, 87.

Klug, A., Finch, J. T., Leberman, R., and Longley, W., 1966*a*, Design and structure of regular virus particles, *Ciba Found. Symp. Princ. Biomol. Organ,* p. 158.

Klug, A., Longley, W., and Leberman, R., 1966*b*, Arrangement of protein subunits and the distribution of nucleic acid in turnip yellow mosaic virus. I. X-ray diffraction studies, *J. Mol. Biol.* **15**, 315.

Kochetkov, N. K., and Budowsky, E. I., 1969, The chemical modification of nucleic acids. *in* "Progress in Nucleic Acid Research" (D. Davidson and W. Cohn eds.), Vol. 9, p. 403, Academic Press, New York.

Kochetkov, N. K., Budowsky, E. I., Sverdlov, E. D., and Symukova, N. A., 1970, "Organic Chemistry of Nucleic Acids," Chimia, Moscow.

Kotaka, T., and Baldwin, R. L., 1964, Effects of nitrous acid on the dAT copolymer as a template for DNA polymerase, *J. Mol. Biol.* **9**, 323.

Krivissky, A. S., Belych, R. A., and Budowsky, E. I., 1973, Mutagenic and inactivating effects of O-methylhydroxylamine on bacteriophage φX174 and its infectious DNA, *Genetika* **9**, 8, 105.

Kuriatkowski, B., Kotarski, J., and Napiorkowska, J., 1973, Photoluminescence studies on the structure of the DNA *in situ* of phages V_i, *Bull. Inst. Med. Mor. Gdansk* **24,** 143.

Kurtz-Fritsch, C., and Hirth, L., 1972, Uncoating of two spherical plant viruses, *Virology* **47,** 385.

Kust, S. V., Dobrov, E. N., and Tikchonenko, T. I., 1972, The investigation of the RNA structure in potato X virus particles, *Mol. Biol.* (*Moscow*) **6,** 42.

Langridge, R., and Gomatos, P. J., 1963, The structure of RNA: Reovirus RNA and transfer RNA have similar three-dimensional structures which differ from DNA, *Science* (*Wash., D.C.*) **141,** 694.

Langridge, R., Seeds, W. E., Wilson, H. R., Hooper, C. W., Wilkins, M. H. F., and Hamilton, L. D., 1957, Molecular structure of deoxyribonucleic acid, *J. Biophys. Biochem. Cytol.* **3,** 767.

Langridge, R., Marvin, D. A., Seeds, W. E., Wilson, H. R., Hooper, C. W., Wilkins, M. H. F., and Hamilton, L. D., 1960, The molecular configuration of deoxyribonucleic acid. II. Molecular models and their Fourier transforms, *J. Mol. Biol.* **2,** 38.

Leach, S. J., and Scheraga, H. A., 1960, Effect of light scattering on ultraviolet difference spectra, *J. Am. Chem. Soc.* **82,** 4790.

Lerman, L. S., 1970, A transition to a compact form of DNA in polymer solutions, *Proc. Natl. Acad. Sci. USA* **68,** 1886.

Lerman, L. S., Jordan, C. F., Venable, J. H., and Maniatis, T. P., 1969, A transition to a compact form of DNA, *Abstr. Third Interntl. Biophys. Congr. Cambridge, Mass.*

Li, H. J., Chang, C., and Weiskopf, M., 1973, Thermal denaturation of nucleohistone effects of formaldehyde reaction, *Biochemistry* **12,** 1763.

Litman, R. M., 1961, Genetic and chemical alterations in the transforming DNA of pneumococcus caused by ultraviolet light and by nitrous acid, *J. Chim. Phys. Phys. Chim. Biol.* **58,** 997.

McCleary, L. O., and Gordon, M. P., 1973, Ultraviolet irradiation of potato virus X, its RNA, and a hybrid virus particle: Photoreactivation, kinetic isotope effects and quantum yield of inactivation, *Photochem. Photobiol.* **18,** 9.

McGavin, S., 1971, Models of specifically paired like (homologous) nucleic acid structures, *J. Mol. Biol.* **55,** 293.

McGavin, S., Wilson, H. R., and Barr, G. C., 1966, Intercalated nucleic acid double helices: A stereochemical possibility, *J. Mol. Biol.* **22,** 187.

Maestre, M. F., 1968, Transient electric birefringence studies of T2 bacteriophage and T2 ghost, *Biopolymers* **6,** 415.

Maestre, M. F., 1970, Circular dichroism of DNA films: Reversibility studies, *J. Mol. Biol.* **52,** 543.

Maestre, M. F., and Kilkson, R., 1962, X-ray investigation of M5 and T2 bacteriophages, *Nature* (*Lond.*) **193,** 366.

Maestre, M. F., and Tinoco, I., Jr., 1967, Optical rotatory dispersion of viruses, *J. Mol. Biol.* **23,** 323.

Maestre, M. F., Gray, D. M., and Cook, R. B., 1971, Magnetic circular dichroism study on synthetic polynucleotides, bacteriophage structure and DNAs, *Biopolymers* **10,** 2537.

Malik, W. U., and Agarwal, S. K., 1967, *p*H-Metric evidence for the binding of magnesium, manganese and strontium with transfusion gelatin, *Ind. J. Chem.* **5,** 1–5.

Manykin, A. A., 1972, Experimental estimation of the time of injection for the T2 phage DNA and analysis of physical mechanism of DNA injection, Ph.D. Thesis, Institute of Virology, Moscow.

Margaretten, W., Morgan, C., Rosenkranz, H. S., and Rose, H. M., 1966, Effect of hydroxyurea on virus development. I. Electron microscopic study of the effect on the development of bacteriophage T4, *J. Bacteriol.* **91,** 823.

Marvin, D. A., 1966, X-ray diffraction and electron microscope studies on the structure of the small filamentous bacteriophage fd, *J. Mol. Biol.* **15,** 8.

Marvin, D. A., and Hohn, B., 1969, Filamentous bacterial viruses, *Bacteriol. Rev.* **33,** 172.

Marvin, D. A., and Schaller, H., 1966, The topology of DNA from the small filamentous bacteriophage fd, *J. Mol. Biol.* **15,** 1.

Marvin, D. A., Spencer, M., Wilkins, M. H. F., and Hamilton, L. D., 1961, The molecular configuration of deoxyribonucleic acid III. X-ray diffraction study of the *C* form of the lithium salt, *J. Mol. Biol.* **3,** 547.

Marvin, D. A., Wiseman, R. L., and Wachtel, E. J., 1974, Filamentous bacterial viruses. XI. Molecular architecture of the class II (pf1, xf) virion, *J. Mol. Biol.* **82,** 121.

Matheka, H. D., Bachrach, H. L., and Trautman, R., 1966, Highly purified foot-and-mouth disease virus: Optical and biological measurements during zone electrophoresis in a glucose density gradient, *Z. Naturforsch.* **21B,** 774.

Mattern, M., Binder, R., and Cerutti, P., 1972, Cytidine photohydration in R17 RNA, *J. Mol. Biol.* **66,** 201–204.

Matthews, K. S., and Cole, R. D., 1972, Shell formation by capsid protein of f2 bacteriophage, *J. Mol. Biol.* **65,** 1.

Matthews, R. E. F., and Ralph, R. K., 1966, Turnip yellow mosaic virus, *Adv. Virus Res.* **12,** 273.

Mayfield, J. E., and Bendet, I. J., 1970*a*, Quantitative flow dichroism. I. Correction for disorientation in a solution of rods, *Biopolymers* **9,** 655.

Mayfield, J. E., and Bendet, I. J., 1970*b*, Quantitative flow dichroism. II. Form dichroism at ultraviolet wavelengths, *Biopolymers* **9,** 669.

Mazurenko, N. N., Budowsky, E. I., and Tikchonenko, T. I., 1972, Early stages of reaction between glyoxal and phage nucleoprotein, *Vopr. Virusol.* **6,** 676.

Mekshenkov, M. I., and Guseynov, R. D., 1971, Interruption of phage T4 chromosome injection into a cell, *Mol. Biol.* (*Moscow*) **5,** 444.

Miall, S. H., and Walker, I. O., 1968, Circular dichroism of *Escherichia coli* ribosomes and TMV, *Biochim. Biophys. Acta* **166,** 711.

Michelson, A. M., 1963, "The Chemistry of Nucleosides and Nucleotides," Academic Press, New York.

Michelson, A. M., Monny, C., and Kapuler, A. M., 1970, Poly-8-bromoguanylic acid, *Biochim. Biophys. Acta* **217,** 7.

Miki, T., and Knight, C. A., 1968, The protein subunit of potato virus X, *Virology* **36,** 168.

Milstein, J. B., and Rossomando, E. T., 1971, Electrooptic studies on the effect of heat treatment on structure in bacteriophage f1, *Virology* **46,** 655.

Ninamishima, Y., Takeya, K., Ohnishi, Y., and Amako, K., 1968, Physicochemical and biological properties of fibrous pseudomonas bacteriophages, *J. Virol.* **2,** 208.

Minchenkova, L. E., Belych, R. A., Dobrov, E. N., and Ivanov, V. I., 1969, Cu^+ and Ag^+ ions use for the investigation of the DNA structure inside phage particles, *Mol. Biol.* (*Moscow*) **3,** 441.

Moll, G., 1963, Elektronenmikroskopische Darstellung des DNA-Fadens im Fortsatz eines Coli-Phagen T2, *Naturwissenschaften* **50,** 411–412.

Moody, M. F., 1965, The shape of the T-even bacteriophage head, *Virology* **26**, 567.
Moore, D. S., and Wagner, T. E., 1973, Origins of the differences between the circular dichroism of DNA and RNA: Theoretical calculations, *Biopolymers* **12**, 201.
Moore, D. S., and Wagner, T. E., 1974, Doublehelical DNA and RNA circular dichroism. Calculations on base-sugar phosphate helix interactions, *Biopolymers* **13**, 977.
Nelson, R. G., and Johnson, W. C., 1970, Conformation of DNA in ethylene glycol, *Biochem. Biophys. Res. Commun.* **41**, 211–216.
North, A. C. T., and Rich, A., 1961, X-ray diffraction studies of bacterial viruses, *Nature (Lond.)* **191**, 1242.
Offord, R. E., 1966, Electron microscopic observations on the substructure of tobacco rattle virus, *J. Mol. Biol.* **17**, 370.
Olins, D. E., and Olins, A. L., 1971, Model nucleohistones: The interaction of F1 and F2al histones with native T7 DNA, *J. Mol. Biol.* **57**, 437.
Olins, D. E., Olins, A. L., and von Hippel, P. H., 1967, Model nucleoprotein complexes: Studies on the interaction of cationic homopolypeptides with DNA, *J. Mol. Biol.* **24**, 157.
Perera, O., and Tikchonenko, T. I., 1969, Study of the phage DNA injection into bacterial cell, *Vopr. Med. Khim.* **15**, 5.
Perham, R. N., and Richards, F. M., 1968, Reactivity and structural role of protein amino groups in TMV, *J. Mol. Biol.* **33**, 795.
Permagorov, V. I., Sladkova, J. A., Velikodvorskaya, G. A., and Tikchonenko, T. I., 1969, Use of dyes for the investigation of structure of DNA in phages, *Mol. Biol. (Moscow)* **3**, 267.
Permagorov, V. I., Debabov, V. G., Sladkova, I. A., and Rebentish, B. A., 1970, Structure of DNA and histones in the nucleohistone, *Biochim. Biophys. Acta* **199**, 556.
Phillips, J. H., Brown, D. M., and Grossman, L., 1966, The efficiency of induction of mutations by hydroxylamine, *J. Mol. Biol.* **21**, 405.
Philipson, K. D., and Sauer, K., 1973, Light-scattering effects on the circular dichroism of chloroplasts, *Biochemistry* **12**, 3454.
Polupanov, V. S., and Cherenkevich, S. N., 1968, Secondary structure of DNA in phage particles, *Biofizika* **13**, 1111.
Pysarevsky, A. N., Gabrilovich, I. M., and Spytkowsky, D. M., 1968, About secondary structure of DNA in phage particles, *Biofizika* **13**, 1101–1113.
Raszka, M., and Mandel, M., 1971, Interaction of aromatic amino acids with neutral polyadenylic acid, *Proc. Natl. Acad. Sci. USA* **68**, 1190.
Rauth, A. M., 1965, Physical state of viral nucleic acid and the sensitivity of viruses to ultraviolet light, *Biophys. J.* **5**, 257.
Reichmann, M. E., 1959, Potato X virus. III. Light scattering studies, *Can. J. Chem.* **37**, 384.
Reichmann, M. E., 1960, Degradation of potato virus X, *J. Biol. Chem.* **235**, 2959.
Remsen, J. F., Miller, N., and Cerutti, P. A., 1970, Photohydration of uridine in the RNA of coliphage R17. II. The relationship between UV inactivation and uridine photohydration, *Proc. Natl. Acad. Sci. USA* **65**, 460.
Remsen, J. F., Mattern, M., Miller, N., and Cerutti, P. A., 1971, Photohydration of uridine in the ribonucleic acid of coliphage R17. Lethality of uridine photohydrates and nonlethality of cyclobutane-type photodimers, *Biochemistry* **10**, 524.
Richards, K. E., Williams, R. C., and Calendar, R., 1973, Mode of DNA packing within bacteriophage heads, *J. Mol. Biol.* **78**, 255.

Rossomando, E. F., and Bladen, H. A., 1969, Physical changes associated with heating bacteriophage f1, *Virology* **39,** 921.

Rossomando, E. F., and Milstein, J. B., 1971, Electro-optic evidence for the control of the structure of bacteriophage f1 by a minor coat protein, *J. Mol. Biol.* **58,** 187.

Rossomando, E. F., and Zinder, N. D., 1968, Studies on the bacteriophage f1. I. Alkali-induced disassembly of the phage into DNA and protein, *J. Mol. Biol.* **36,** 387.

Rubinstein, I., 1960, Ph.D. Thesis, University of California, Los Angeles, Calif. (quoted by Kilkson and Maestre, 1962).

Rushizky, G. W., Knight, C. A., and McLaren, A. D., 1960, A comparison of the ultraviolet-light inactivation of infectious ribonucleic acid preparations from tobacco mosaic virus with those of the native and reconstituted virus, *Virology* **12,** 32.

Rvachev, V. P., Sachnovsky, M. Y., Gumenetsky, S. G., Tikchonenko, T. I., and Dobrov, E. N., 1968, Study of absorbancy of viral suspension with integrating photometer, *Zh. Prikl. Spektrosk.* **8,** 844.

Schachter, E. M., Bendet, I. R., and Lauffer, M. A., 1966, Orientation of the RNA in tobacco mosaic virus, *J. Mol. Biol.* **22,** 165.

Schauenstein, E., and Bayzer, H., 1955, Über die quantitative Berücksichtigung der Tyndall-Absorption im UV-Absorptions-spektrum von Proteinen, *J. Polymer Sci.* **16,** 45–52.

Schellman, J. A., and Schellman, C., 1966, The conformation of polypeptide chains in proteins, *in* "Proteins" (H. Neurath, ed.) Vol. 2, p. 1, Academic Press, New York.

Schramm, G., and Zillig, W., 1955, Über die Struktur des Tabakmosaikvirus. IV. Die Reaggregation des nucleinsäurefreien Proteins, *Z. Naturforsch.* **10B,** 493.

Schubert, D., and Frank, H., 1971, Properties of particles aggregated from protein subunits of bacteriophage fr, *Virology* **43,** 41.

Schuster, H., and Vielmetter, W., 1961, Studies on the inactivating and mutagenic effect of nitrous acid and hydroxylamine on viruses, *J. Chim. Phys. Physochim. Biol.* **58,** 1005–1010.

Schuster, H., and Wilhelm, R. C., 1963, Reaction differences between TMV and its free ribonucleic acid with nitrous acid, *Biochim. Biophys. Acta* **68,** 554.

Sehgal, O. P., 1973, Inactivation of southern bean mosaic virus and its ribonucleic acid by nitrous acid and ultraviolet light, *J. Gen. Virol.* **18,** 1.

Sehgal, O. P., and Krause, G. F., 1968, Efficiency of nitrous acid as an inactivating and mutagenic agent of intact tobacco mosaic virus and its isolated nucleic acid, *J. Virol.* **2,** 966.

Sehgal, O. P., and Soong, M. M., 1972, Reaction of nitrous acid with viral nucleic acids *in situ, Virology* **47,** 239.

Sellini, H., Maurizot, J. C., Dimicoli, J. L., and Hélène, C., 1973, Hydrogen bonding of amino acid side chains to nucleic acid bases, *FEBS (Fed. Eur. Biochem. Soc.) Lett.* **30,** 219.

Semenov, M. A., Gasan, A. I., and Maleev, V. Y., 1971, Study of thermal destruction of T2 phage and its components with infra-red spectroscopy and adiabatic calorimetry, *Dokl. Akad. Nauk SSSR* **198,** 1449.

Shapiro, J. T., Leng, M., and Felsenfeld, G., 1969, Deoxyribonucleic acid–polylysine complexes. Structure and nucleotide specificity, *Biochemistry* **8,** 3219.

Shapiro, R., Cohen, B. I., and Clagett, D. C., 1970, Specific acylation of the guanine residues of ribonucleic acid, *J. Biol. Chem.* **245,** 2633.

Shapiro, R., and Pohl, S. H., 1968, The reaction of ribonucleosides with nitrous acid. Side products and kinetics, *Biochemistry* **7,** 448.

Shatsky, I. N., Chichkova, N. V., and Bogdanov, A. A., 1971, RNA-protein interactions in the ribosomes, *Mol. Biol.* (*Moscow*) **5,** 817.

Shepherd, R. J., Wakeman, R. J., and Ghabrial, S. A., 1968, Preparation and properties of the protein and nucleic acid components of pea enation mosaic virus, *Virology* **35,** 255.

Shie, M., Chirgadze, Y. N., and Tikchonenko, T. I., 1970, A study of free and intraphage DNA hydration using the infra-red spectroscopy, *Vopr. Virusol.* **5,** 619.

Shie, M., Kharitonenkov, I. G., Tikchonenko, T. I., and Chirgadze, Y. N., 1972*a*, New possibilities of investigating nucleic acids and nucleoproteins in aqueous solutions by infrared spectroscopy, *Nature* (*Lond.*) **235,** 386.

Shie, M., Nevskaya, N. A., and Chirgadze, Y. N., 1972*b*, The infrared spectra of water solutions of nucleic acids and the nucleoprotein complexes in the region of sugar-phosphate skleleton vibrations. *Abst. IV Interntl. Biophys. Congr. Moscow* **CVII.**

Shih, T. Y., and Fasman, G. D., 1971, Circular dichroism studies of deoxyribonucleic acid complexes with arginine-rich histone IV (f2al), *Biochemistry* **10,** 1675.

Shih, T. Y., and Fasman, G. D., 1972, Circular dichroism studies of histone-deoxyribonucleic acid complexes. A comparison of complexes with histone I (f-1), histone IV (f2a1), and their mixtures, *Biochemistry* **11,** 398.

Siegel, A., and Norman, A., 1958, Action spectra for two strains of tobacco mosaic virus, *Virology* **6,** 725.

Siegel, A., Wildman, S. G., and Ginoza, W., 1958, Sensitivity to ultra-violet light of infectious TMV nucleic acid, *Nature* (*Lond.*) **178,** 1117.

Simmons, N. S., and Blout, E. R., 1960, The structure of TMV and its components: Ultraviolet optical rotatory dispersion, *Biophys. J.* **I,** 55.

Simmons, N. S., and Glazer, A. N., 1966, Reversible disorientation of RNA bases in tobacco mosaic virus (TMV). Optical rotatory dispersion, *Abstr. II Interntl. Biophys. Congr. Vienna,* p. 208.

Simon, L. D., and Anderson, T. F., 1967, The infection of *Escherichia coli* by T2 and T4 bacteriophages as seen in the electron microscope. I. Attachment and penetration, *Virology* **2,** 279.

Simukova, N. A., and Budowsky, E. I., 1974, Conversion of non-covalent interactions in nucleoproteins into covalent bonds: UV-induced formation of polynucleotide-protein crosslinks in bacteriophage Sd virions, *FEBS* (*Fed. Eur. Biochem. Soc.*) *Lett.* **38,** 299.

Singer, B., 1971, Chemical modification of viral ribonucleic acid. IX. The effect of ultraviolet irradiation on TMV-RNA and other polynucleotides, *Virology* **45,** 101.

Singer, B., and Fraenkel-Conrat, H., 1969*a*, Chemical modification of viral ribonucleic acid. VII. The action of methylating agents and nitrosoguanidine on polynucleotides including TMV ribonucleic acid, *Biochemistry* **8,** 3260.

Singer, B., and Fraenkel-Conrat, H., 1969*b*, Chemical modification of viral ribonucleic acid. VIII. The chemical and biological effects of methylating agents and nitrosoguanidine on tobacco mosaic virus, *Biochemistry* **8,** 3266.

Singer, B., and Fraenkel-Conrat, 1969*c*, Mutagenicity of alkyl and nitroso-alkyl compounds acting on tobacco mosaic virus and its RNA, *Virology* **39,** 395.

Singer, B., and Fraenkel-Conrat, H., 1970, Messenger and template activities of chemically modified polynucleotides, *Biochemistry* **9,** 3694.

Sinha, R. K., and Misra, D. N., 1971, Studies on the secondary structure of intraphage T-7 DNA, *Z. Naturforsch.* **26B,** 1288.

Sinsheimer, R. L., 1959, Purification and properties of bacteriophage φX174, *J. Mol. Biol.* **1,** 37.

Skladneva, V. B., Budowsky, E. I., and Tikchonenko, T. I., 1973, Study of modification of Sd phage by bisulfite, *in* "Molecular Biology of Viruses," p. 86, Academy of Medical Sciences & Institute of Virology, Moscow.

Smith, K. C., and Aplin, R. T., 1966, A mixed photoproduct of uracil and cysteine (5-*S*-cysteine-6-hydrouracil). A possible model for the *in vivo* cross-linking of deoxyribonucleic acid and protein by ultraviolet light, *Biochemistry* **5,** 2125.

Smith, K. C., and Meun, D. H. C., 1968, Kinetics of the photochemical addition of [^{35}S] cysteine to polynucleotides and nucleic acids, *Biochemistry* **7,** 1033.

Smith, K. C., and O'Leary, M. E., 1967, Photoinduced DNA–protein cross-links and bacterial killing: A correlation at low temperatures, *Science (Wash., D.C.)* **155,** 1024.

Smith, K. C., Hodgkins, B., and O'Leary, M. E., 1966, The biological importance of ultraviolet light induced DNA–protein crosslinks in *Escherichia coli*$_{15}$TAU, *Biochim. Biophys. Acta* **114,** 1.

Solari, A. J., 1965, Structure of the chromatin in sea urchin sperm, *Proc. Natl. Acad. Sci. USA* **53,** 503.

Spirin, A. S., 1963, Some problems concerning the macromolecular structure of ribonucleic acids, *Progr. Nucleic Acid Res.* **I,** 30.

Spitkowsky, D. M., Andrianov, V. T., and Pisarevsky, A. T., 1969, "Radiation Biophysics of Nucleoproteins," Atomizdat, Moscow (in Russian).

Staehelin, M., 1957, Inactivation of TMV-RNA with formaldehyde, *Fed. Proc.* **16,** 254.

Staehelin, M., 1958, Reaction of TMV nucleic acid with formaldehyde, *Biochim. Biophys. Acta* **29,** 410.

Starowsky, O. V., 1971, An approach to the calculation of permeability of the TMV capsid, *in* "Voprosy Obshchey Virusologii," Vol. 1, p. 44, Institute of Virology, Academy of Medical Sciences, Moscow.

Stols, A. L. H., and Veldstra, H., 1965, Interactions of turnip yellow mosaic virus with quaternary ammonium salts, *Virology* **25,** 508.

Streeter, D. G., and Gordon, M. P., 1968, A study of inactivation "and reactivation" in the UV irradiated TMV and TMV RNA, *Photochem. Photobiol.* **8,** 81.

Stroke, G. W., and Haliova, M., 1972, Attainment of diffraction limited in high-resolution electron microscopy by a *posterioli* holographic image sharpening, *Optik* **35,** 50–65.

Sundaralingham, M., 1969, Stereochemistry of nucleic acids and their constituents. IV. Allowed and preferred conformations of nucleosides, nucleoside mono-, di-, tri-, tetraphosphates, nucleic acids and polynucleotides, *Biopolymers* **7,** 821.

Sutherland, G. B. B. M., and Tsuboi, M., 1957, The infra-red spectrum and molecular configuration of sodium deoxyribonucleate, *Proc. R. Soc. Lond. Ser. A Math. Phys. Sci.* **239,** 446.

Suwalsky, M., and Traub, W., 1972, A comparative X-ray study of a nucleoprotamine and DNA complexes with polylysine and polyarginine, *Biopolymers* **11,** 2223.

Takeya, K., and Amako, K., 1966, A rod-shaped pseudomonas phage, *Virology* **28,** 163.

Taniguchi, M., Yamaguchi, A., and Taniguchi, T., 1971, Flow dichroic spectra of TMV and their protein assemblies, *Biochim. Biophys. Acta* **251,** 164.

Tao, M., Gordon, M. P., and Nester, E. W., 1966, Kinetic isotope studies on the inactivation of transforming deoxyribonucleic acid, tobacco mosaic virus, and its nucleic acid by ultraviolet light, *Biochemistry* **5,** 4146.

Tao, M., Small, G., O'Brien, L., and Gordon, M. P., 1968, Photochemical alterations of TMV RNA, *Fifth Interntl. Congr. Photobiol. Hanover, New Hampshire*, p. 109.
Tao, M., Small, G. D., and Gordon, M. P., 1969, Photochemical alterations in ribonucleic acid isolated from ultraviolet-irradiated tobacco mosaic virus, *Virology* **39,** 534.
Thomas, C. A., 1963, The organization of DNA in bacteriophage and bacteria, *in* "Molecular Genetics" (J. H. Taylor, ed.), p. 113, Academic Press, New York.
Tikchonenko, T. I., 1969, Conformation of viral nucleic acids *in situ*, *Adv. Virus Res.* **15,** 201.
Tikchonenko, T. I., 1971, About the secondary structure of DNA in virus particles, *Vestn. Akad. Med. Nauk SSSR* **2,** 46.
Tikchonenko, T. I., 1974, The DNA-protein interactions in the phage nucleoproteins, *in* "Structure and Functions of Nucleic Acids and Nucleoproteins," Thesis of N. A. Belozetsky Symposium, Jan. 29–Feb. 2, 1974, Moscow State University, Moscow.
Tikchonenko, T. I., and Dobrov, E. N., 1969, Peculiarities of the secondary structure of bacteriophage *in situ*. II. Reaction with formaldehyde, *J. Mol. Biol.* **42,** 119.
Tikchonenko, T. I., Dobrov, E. N., Velikodvorskaya, G. A., and Kisseleva, N. P., 1966, Peculiarities of the secondary structure of phage DNA *in situ*, *J. Mol. Biol.* **18,** 58.
Tikchonenko, T. I., Budowsky, E. I., Sklyadneva, V. B., and Khromov, I. S., 1971, The secondary structure of bacteriophage DNA *in situ*. III. Reaction of Sd phage with O-methylhydroxylamine, *J. Mol. Biol.* **55,** 535.
Tikchonenko, T. I., Kisseleva, N. P., Zintshenko, A. I., Ulanov, B. P., and Budowsky, E. I., 1973, Peculiarities of the secondary structure of bacteriophage DNA *in situ*. IV. Covalent cross-links between DNA and protein that arise in the reaction of Sd phage with O-methylhydroxylamine, *J. Mol. Biol.* **73,** 109.
Tikchonenko, T. I., Kislina, O. S., and Dobrov, E. N., 1974*a*, Peculiarities of the secondary structure of bacteriophage DNA *in situ*. V. Change in DNA conformation inside the phages under the influence of formaldehyde, *Arch. Biochem. Biophys.* **160,** 1.
Tikchonenko, T. I., Budowsky, E. I., and Mazurenko, N. N., 1974*b*, Peculiarities of the secondary structure of bacteriophage DNA *in situ*. VI. The reaction of phage Sd with glyoxal (in press).
Tikchonenko, T. I., Andronikova, M. L., Tchruni, F. I., and Kisseleva, N. P., 1974*c*, Manuscript in preparation.
Tinoco, I., Jr., 1960, Hypochromism in polynucleotides, *J. Am. Chem. Soc.* **82,** 4785–4790.
Tollin, P., Wilson, H. R., Young, D. W., Cathro, J., and Mowat, W. P., 1967, X-ray diffraction and electron microscope studies of narcissus mosaic virus, and comparison with potato virus X, *J. Mol. Biol.* **26,** 353.
Tomita, K.-I., and Rich, A., 1964, X-ray diffraction investigations of complementary RNA, *Nature* (*Lond.*) **201,** 1160.
Tramer, Z., Wierzchowski, K. L., and Shugar, D., 1969, Influence of polynucleotide secondary structure on thymine photodimerization, *Acta Biochim. Pol.* **16,** 83.
Travers, F., Michelson, A. M., and Douzou, P., 1970, Conformational changes of nucleic acids in methanol-water solutions at low temperature, *Biochim. Biophys. Acta* **217,** 1.
Tremaine, J. H., and Goldsack, D. E., 1968, The structure of regular viruses in relation to their subunit amino acid composition, *Virology* **35,** 227.
Tsuboi, M., 1969, Application of infrared spectroscopy to structure studies of nucleic acids, *Appl. Spectr. Rev.* **3,** 45.

Tsugita, A., and Fraenkel-Conrat, H., 1963, Contributions from TMV studies to the problem of genetic information transfer and coding., *in* "Molecular Genetics" (J. H. Taylor, ed.), Part 1, Chapt. X, p. 477, Academic Press, New York.

Tung, J-S., and Knight, C. A., 1972, The coat protein subunits of cucumber viruses 3 and 4 and a comparison of methods for determining their molecular weights, *Virology* **48,** 574.

Tunis, M. J., and Hearst, J. E., 1968, Optical rotatory dispersion of DNA in concentrated salt solutions, *Biopolymers* **6,** 1218–1223.

Tunis-Schneider, M. J., and Maestre, M. F., 1970, Circular dichroism spectra of oriented and unoriented deoxyribonucleic acid films—A preliminary study, *J. Mol. Biol.* **52,** 521.

Turchinsky, M. F., Kusova, K. S., and Budowsky, E. I., 1974, Conversion of non-covalent interactions in nucleoproteins into covalent bonds: Bisulfide-induced formation of polynucleotide protein crosslinks in MS2 bacteriophage virions, *FEBS* (*Fed. Eur. Biochem. Soc.*) *Lett.* **38,** 304.

Vainstein, B. K., 1963, "X-ray Diffraction by Chain Molecules," Nauka, Moscow.

van Kammen, A., 1972, Plant viruses with a divided genome, *Annu. Rev. Plant Pathol.* **10,** 125.

van de Hulst, H. C., 1957, "Light Scattering by Small Particles," John Wiley & Sons, New York.

Varma, A., Gibbs, A. J., Woods, R. D., and Finch, J. T., 1968, Some observations on the structure of the filamentous particles of several plant viruses, *J. Gen. Virol.* **2,** 107.

Varshavsky, Y. M., Evdokimov, Y. M., and Akimenko, N. M., 1973, A compact form of double-stranded DNA in solution, *Stud. Biophys.* **40,** 41–56.

Velikodvorskaya, G. A., Klimenko, S. M., Mazzarelli, M., and Tikchonenko, T. I., 1968, Interaction of phages with bacterial cell walls, *Mol. Biol.* (*Moscow*) **2,** 519.

Vielmetter, W., and Schuster, H., 1960, The base specificity of mutation induced by nitrous acid in phage T2, *Biochem. Biophys. Res. Commun.* **2,** 324.

von Hippel, P. H., and McGhee, J. D., 1972, DNA-protein interactions, *Annu. Rev. Biochem.* **41,** 231.

Wagner, K. G., and Arav, R., 1968, On the interaction of nucleotides with poly-L-lysine and poly-L-arginine. I. The influence of the nucleotide base on the binding behavior, *Biochemistry* **7,** 1771.

Wang, T. M., and McLaren, A. D., 1972, Conformational changes induced in tobacco mosaic virus nucleic acid by ultraviolet radiation, *Biophysik* **8,** 237.

Warshaw, M. M., and Tinoco, I., Jr., 1965, Absorption and optical rotatory dispersion of six dinucleoside phosphates, *J. Mol. Biol.* **13,** 54.

Werbin, H., Valentine, R. C., and McLaren, A. D., 1967, Photobiology of RNA bacteriophages. I. Ultraviolet inactivation and photoreactivation studies, *Photochem. Photobiol.* **6,** 205–213.

Werbin, H., Valentine, R. C., Hidalgo-Salvatierra, O., and McLaren, A. D., 1968, Photobiology of the RNA bacteriophages. II. UV-irradiation of f2: Effects on extracellular stages of infection and on early replication, *Photochem. Photobiol.* **7,** 253–261.

Wetlaufer, D. B., 1962, Ultraviolet spectra of proteins and amino acids, *Adv. Protein Chem.* **17,** 303.

White, R. A., and Fischbach, F. A., 1973, An X-ray scattering investigation of broad bean mottle virus in solutions of various electron densities, *J. Mol. Biol.* **75,** 549.

Wilson, H. R., and Tollin, P., 1969, Some observations on the structure of potato virus X, *J. Gen. Virol.* **5,** 151.

Wiseman, R. L., Dunker, A. K., and Marvin, D. A., 1972, Filamentous bacterial viruses. III. Physical and chemical characterization of the Ifl virion, *Virology* **48,** 230.

Yamada, Y., Shigeta, A., and Nozu, K., 1973, Ultraviolet effects on biological function of RNA phage MS2, *Biochim. Biophys. Acta* **299,** 121.

Yang, J. T., and Samejima, T., 1969, Optical rotatory dispersion and circular dichroism of nucleic acids, *Progr. Nucleic Acid Res. Mol. Biol.* **9,** 223.

Zaretsky, I. Z., Farashyan, V. P., and Tikchonenko, T. I., 1971, A study of kinetics of T2 phage adsorption on *E. coli* B cells and stabilization of phage-cell complex against treatment in homogenizer, *Vopr. Med. Khim.* **17,** 315.

Zarybnicky, V., 1969, Mechanism of T-even DNA ejection, *J. Theor. Biol.* **22,** 33.

Zinder, N. D., 1965, RNA phages, *Annu. Rev. Microbiol.* **19,** 455.

Zipper, P., Kratky, O., Herrmann, R., and Hohn, T., 1971, An X-ray small angle study of the bacteriophages fr and R17, *Eur. J. Biochem.* **18,** 1.

Zubay, G., and Wilkins, M. H. F., 1960, X-ray diffraction studies of the structure of ribosomes from *Escherichia coli, J. Mol. Biol.* **2,** 105.

CHAPTER 2

Structure of Small DNA-Containing Animal Viruses

John T. Finch
MRC Laboratory of Molecular Biology
Cambridge, United Kingdom

and

Lionel V. Crawford
Imperial Cancer Research Fund Laboratories
London, United Kingdom

1. GENERAL PRINCIPLES OF VIRUS STRUCTURE

One group of small DNA-containing animal parvoviruses has already been dealt with in a recent chapter (Rose, this series, Volume 4). The two groups of viruses which come next in size, the polyoma viruses and papilloma viruses, have attracted a great deal of attention especially in recent years. The properties and suggested nomenclature for viruses in these groups are covered in a recent publication (Melnick *et al.*, 1974). In this chapter we will review the present state of knowledge regarding the chemical and physical make-up of these viruses.

The general principles involved in the construction of the smaller regular viruses will be described in detail by Caspar in a later volume of this series. These ideas have evolved from consideration of the structures of the simplest viruses, in which the infectious nucleic acid is contained within a protein shell, or capsid, built from many identical protein subunits. In forming the capsid, the identical subunits tend to associate with their neighbors in an identical fashion and thus generate a repeating pattern over the shell which can be described in terms of symmetry operations. In the case of isometric shells, the largest number of subunits that can be accommodated so that each is in a strictly equivalent environment is 60, arranged with the symmetry of the icosahedral point group, 532. Caspar and Klug (1962) showed that by introducing further local symmetry in addition to that of the 532 point group, it would be possible to accommodate larger numbers of subunits which are in very nearly (or quasi-) equivalent environments. These possible extra local symmetries were classified in terms of a triangulation number T which can take the values 1, 3, 4, 7, 9, etc., specifying icosahedral surface lattices which can accommodate $60T$ structure units (the identical protein molecules). The icosahedral surface lattice specified by a particular value of T is obtained by subtriangulating the surface of a sphere into $20T$ triangles, consistent with icosahedral symmetry.

This method of classification has proved invaluable in determining the surface structure of more complex viruses. Thus, although it is not unlikely that the capsids of polyoma and the papilloma viruses are built from more than one protein, the surface patterns fall into the above classification scheme. By postulating particular positions of the subunits of a given lattice, consistent with the symmetry (both local and strict) of the lattice, the appearance of a capsid constructed according to the principles of quasiequivalence can be built up. Certain striking surface patterns are generated by subunits located close to symmetry axes so that they cluster in well-defined groups which are easily resolved by the electron microscope. In particular, the large morphological units seen on the surfaces of the polyoma and papilloma viruses are consistent with the hexamer and pentamer clustering of units at the vertices of the surface lattice at which are located the local sixfold and fivefold rotational symmetry axes.

2. PAPILLOMA VIRUSES

2.1. Biology of the Viruses*

Papilloma viruses cause benign skin tumors in a variety of mammalian species. These papillomas contain relatively large amounts of

* See N. P. Salzman and G. Khoury, Chapt. 2, Vol. 3 of this series for a more detailed discussion.

virus and the extraction and purification of the virus from papilloma tissue is straightforward. The papilloma tissue, either fresh or stored in glycerol, is ground with sand or homogenized. After the cell debris has been removed by low-speed centrifugation, the virus can be spun down at high speed. For many purposes the virus is sufficiently pure to be used at this stage simply because there is far less nonviral material in papilloma tissue than in other sources. Further purification by equilibrium density gradient centrifugation removes residual traces of cellular material and allows DNA-containing particles to be separated from particles which lack DNA. Because of the high yield and stability of virus and ease of extraction and purification, these viruses were among the first animal viruses to become available for detailed physical and structural studies. The virus structure is now known in considerable detail, although, because no suitable tissue-culture system for virus propagation has been developed, little is known about their replication. Viruses of this group differ from most other DNA tumor viruses in several respects. First, they cause tumors in their natural hosts. The papilloma virus of each mammalian species appears to be distinct, both antigenically (Le Bouvier *et al.*, 1966) and with respect to the base composition of the DNA (Crawford and Crawford, 1963), and each seems to be restricted to a particular species. Second, the tumors themselves contain substantial quantities of virus. This is in contrast to those tumors caused by other DNA tumor viruses which usually contain little or no detectable virus. The viral DNA replicates in the proliferating layer of the dermis and virus particles are only produced as the cells move outward and become keratinized.

2.2. Composition

2.2.1. DNA Content and Molecular Mass

Papilloma virus particles are composed of an outer protein shell with icosahedral symmetry enclosing a molecule of DNA. The virus DNA is double-stranded and has a covalently closed circular structure. When extracted from the virus the DNA takes up a twisted, superhelical configuration due to the defiency of turns in the DNA double helix. The molecular mass of the DNA is 5×10^6 daltons and that of the virus particle 40×10^6 daltons (Watson and Littlefield, 1960; Crawford and Crawford, 1963; Crawford, 1965; Kleinschmidt *et al.*, 1965). The DNA content of 12% wt./wt., by direct chemical estimation (Kass and Knight,

1965), and the density of the virus particle, 1.34 g/ml, are consistent with these values. Probably the viral DNA is complexed with an equal weight of internal protein, as is the case for polyoma virus and SV40 (see below). This leaves 30×10^6 daltons of coat protein. The simplest structure consistent with the morphology of the particles would then be made up of 420 polypeptide subunits each with a molecular mass of about 70,000 daltons ($30 \times 10^6/420$).

2.2.2. Polypeptides

Although papilloma viruses were among the first tumor viruses to be isolated, there was, until recently, no information on their polypeptide composition. The amino acid composition of Shope papilloma virus was determined by Kass and Knight (1965) and showed no unusual features. The antigenic properties of polypeptides from human wart virus were examined by Pass and Maizel (1973). They showed that the major virus particle polypeptide had a molecular mass of 53,000 daltons and that sera from some patients with warts reacted with this polypeptide. In addition full particles contained three small polypeptides which were absent from empty particles. These were probably the same polypeptides, corresponding to cellular histones, which were observed by Spira *et al.* (1974) in a more extensive comparison of SV40, human wart, and Shope rabbit papilloma virus proteins. Their results are summarized in Table 1. The nomenclature used for the polypeptides here is awkward in that the major capsid protein is VP3 for rabbit papilloma virus and VP4 for human papilloma virus, in contrast to polyoma and SV40, where VP1 is the major capsid protein. The molecular weights determined for the major capsid proteins of the two papilloma viruses were different, 60,000 and 63,000. The latter value for human papilloma virus is not in agreement with the value of 53,000 mentioned previously (Pass and Maizel, 1973). Our own determinations on human papilloma virus gave a value of 62,000 for the major capsid protein, close to that obtained by Spira *et al.*, (1974). Iodination of the intact virus particle failed to label the major capsid proteins of either human or rabbit papilloma viruses, in contrast to the results with SV40 (Spira *et al.*, 1974), and polyoma (Gibson 1974). This is unusual and must indicate that in the intact particle, tyrosine residues are not exposed, although the rabbit papilloma virus particle does contain tyrosine (Kass and Knight, 1965) and it seems likely that much of it is in the major capsid protein.

TABLE 1

Molecular Weights of Papilloma Virus Particle Polypeptides[a]

Virus		Molecular weight	Percentage of total protein
SV40	VP1	43,000	69
	VP2	32,000	4
	VP3	24,000	14
	VP4–5	14,000	11
	VP6	12,000	1
Rabbit papilloma	VP1	85,000	11
	VP2	70,000	21
	VP3	60,000	48
	VP4	19,000	9
	VP5	15,000	7
Human papilloma	VP1	115,000	8
	VP2	100,000	1
	VP3	85,000	9
	VP4	63,000	60
	VP5	51,000	17
	VP6	14,000	1

[a] The major virion polypeptide is underlined in each case. Data from Spira *et al.*, 1974.

2.3. Structure

2.3.1. Historical Summary

Information on the physical structure of the papilloma viruses has been obtained exclusively by electron microscopy. The difficulties of studying crystals of viruses of this size in detail by X-ray diffraction have not yet been overcome, and the amount of papilloma virus available has not been sufficient to allow the investigation of the optimum conditions of crystallization. Thus, although crystals of human papilloma virus were grown (Finch and Leberman, unpublished work), they were of poor shape and X-ray diffraction showed them to be of poor order.

Early electron micrographs by R. C. Williams (1953) of air-dried, metal-shadowed rabbit papilloma virus showed that the virus particles were isometric and about 50 nm in diameter. These were the first electron micrographs of a virus to show substructure. Protuberances spaced about 7 nm apart were observed in a close-packed array over the surface of the virus particle. In later studies by R. C. Williams *et al.* (1960) the number of protuberances was counted and found to be about 30 on the shadowed halves of particles; a similar number of surface knobs was seen in the images of negatively stained particles. The same conclusion was reached by Breedis *et al.* (1962), but neither of these sets of authors proposed a definite structural model beyond concluding that the surface of the particle was composed of 60–70 of these large, morphological units.

A definite model consisting of 42 morphological units, however, was proposed by M. G. Williams *et al.* (1961) for human papilloma virus, but no critical analysis of the images was presented (Klug, 1965). Detailed analysis of the images, however, was later presented for the human and the rabbit papilloma viruses (Klug and Finch, 1965; Finch and Klug, 1965), and this showed that both viruses had 72 morphological unit structures, although of opposite handedness, as described in Sect. 2.3.2. This result was later confirmed by tilting experiments in the electron microscope and by the three-dimensional image reconstruction of human papilloma virus, also described in Sect. 2.3.2.

2.3.2. Morphology

Crude virus preparations are easily prepared as described above and examination of such a crude extract with the electron microscope shows a variety of structures. Thus, in the field shown in Fig. 1, which is of a negatively stained crude extract of rabbit papilloma virus, the dominant particles are isometric about 50 nm in diameter, some of which are penetrated by the stain and thus appear hollow. However, there are also elongated tubular structures of various sorts and occasional smaller isometric particles. All of these, however, possess similar surface detail, consisting of large morphological units which are locally close packed, about 10 nm apart. Also visible in the field is an array of more than 60 small polygonal units which has evidently arisen by the disintegration of one of the larger particles into the constituent morphological units which have settled onto the specimen film.

The centrifugation of such a crude virus preparation in a density gradient yields two main bands. The band of greater density yields infective virus particles, and electron micrographs show them to be isometric in shape (Fig. 2). The diameters of the particle images vary with the degree of preservation and of staining, but in hexagonally close-packed arrays the interparticle distance, which may be taken as a maximum diameter, is about 56 nm (Klug and Finch, 1965). A similar value was obtained for the interparticle distance in the three-dimen-

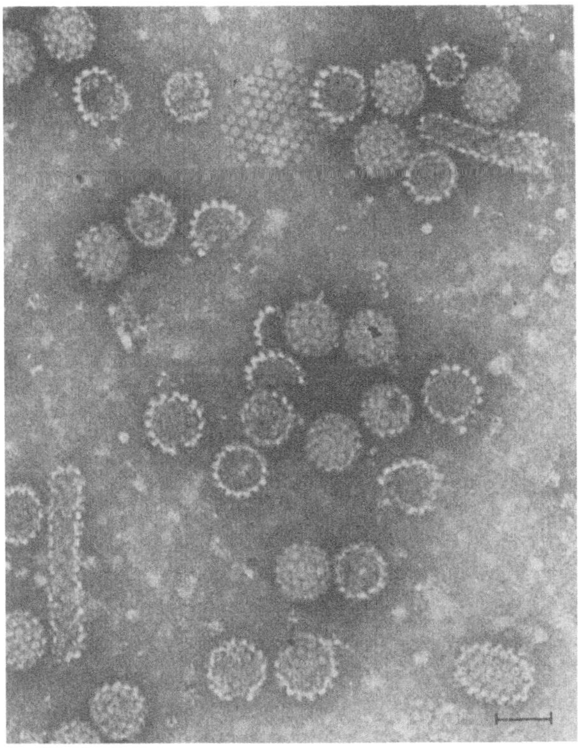

Fig. 1. Electron micrograph of a negatively stained preparation of unfractionated rabbit papilloma virus (bar = 50 nm). In addition to the uniform circular images of intact virus particles about 50 nm in diameter, the preparation shows empty penetrated particles, tubes, elongated particles, an abnormally small circular particle, and a puddle of morphological units, presumably from a disrupted shell, which have formed into a regular planar array.

sional crystals of human papilloma virus referred to earlier. The main bodies of the virus particles are usually unpenetrated by stain.

The other particles observed in unfractionated extracts form the lighter, noninfectious protein band in the density gradient. Unlike the virus particles, these are usually penetrated by stain and so present a hollow appearance in electron micrographs.

2.3.3. Interpretation of Surface Structure

The electron microscope images of both negatively stained and shadowed particles of the papilloma viruses show large morphological

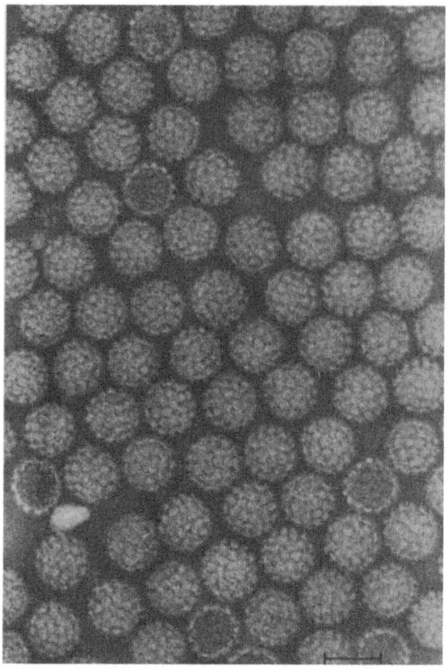

Fig. 2. Electron micrograph of negatively stained, purified human papilloma virus (bar = 50 nm). Apart from a few particles which have been penetrated by stain, the images are uniform in appearance, about 50 nm in diameter.

units which are locally arranged in hexagonal close packing. As described above, the number of morphological units over the surface was estimated by Williams *et al.* (1960) and Breedis *et al.* (1962) to be 60–70. On the basis of the theory of Caspar and Klug (1962) one would expect that these large units result from the individual protein structure units clustering together in hexamers and pentamers at the vertices of an icosahedral surface lattice. For the lattice specified by the triangulation number T, such clustering would yield $(10T + 2)$ morphological units made up of $10(T - 1)$ hexamers and 12 pentamers. Thus, the estimate of 60–70 strongly suggests the presence of the $T = 7$ lattice. Two representations of this lattice are shown in Fig. 3, and it is evident that it is skew (i.e., possesses no mirror planes of symmetry) and so can exist in both a left- and right-handed form, the $T = 7l$ and $T = 7d$ lattices, respectively. Both representations in Fig. 3 are of the $T = 7d$

lattice, the path between neighboring 5-coordinated lattice points (vertices) being like that of a knight's move in chess, 2 lattice steps out and one to the right (5—6—6↘5). The corresponding $T = 7l$ lattices are the mirror images of these, and the path between the 5-vertices would be two steps out and one to the left (5—6—6↗5). Models showing the 72 morphological unit structures arising from the clustering of structure units into hexamers and pentamers are shown in Fig. 4.

In order to prove from isolated electron microscope images that the papilloma surface structure is indeed based on the $T = 7$ icosahedral surface lattice, it is necessary to analyze the arrangement of

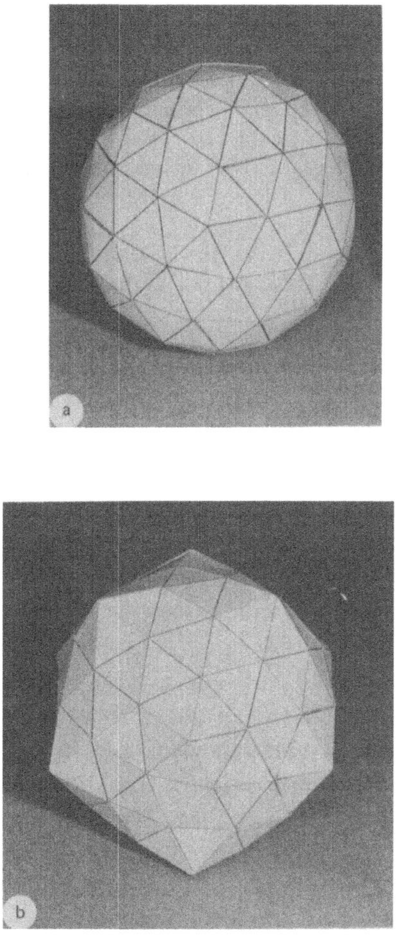

Fig. 3. Two representations of the icosahedral surface lattice $T = 7d$. (a) The geodesic sphere in which all the vertices are at the same radius. Units arranged on this lattice would be closer together near the fivefold vertices than elsewhere since the edges meeting at these vertices are shorter than the rest. (b) The deltahedron, in which all the edge lengths are the same but the vertices are not all at the same radius. Units arranged on this lattice would have the same surface density near the fivefold vertices as elsewhere but would lie at a greater radius. The corresponding $T = 7l$ lattices are the mirror images of a and b.

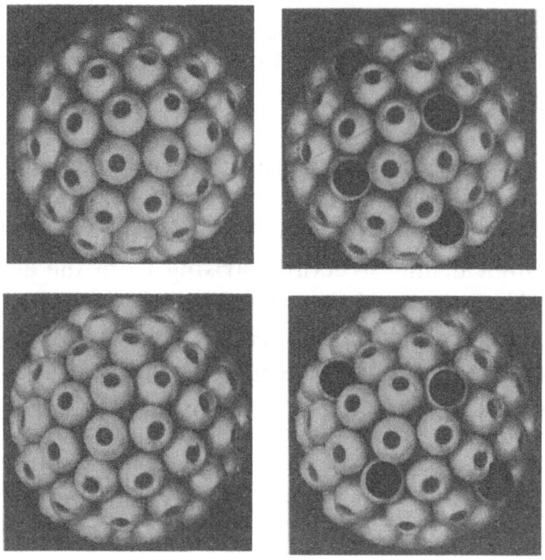

T = 7l

T = 7d

Fig. 4. Models of icosahedral shells with 72 morphological units at the vertices of the icosahedral surface lattices $T = 7l$ and $T = 7d$. The 5-coordinated units are marked in the right-hand copies. The paths between these fivefold units are two steps out and one to the left (5—6—6↗5) for the $T = 7l$ arrangement and two steps out and one to the right (5—6—6↘5) for the $T = 7d$.

hexamers between neighboring pentamers. Pentamers can be recognized as those morphological units which are surrounded by five-nearest-neighbor units; they are shown black in the models in Fig. 4b. This method of analysis is particularly difficult in the case of the $T = 7$ lattice because of its skew nature. The method of specimen preparation which yields images of virus particles showing the finest detail is that of negative staining—outlining the particles in an electron-dense stain. In general, the stain coats the complete surface of the virus, although maybe to varying extents, so that the resultant image obtained is a superposition of detail from both the near and far sides of the particle. With viruses based on symmetrical lattices, for views normal to mirror planes, both sides of the lattice superpose and usually yield striking and informative images. This cannot occur for an arrangement based on a skew lattice and so inevitably in these images there is always at least some interleaving of detail from opposite sides of the particle, thus confusing an analysis based on one side of its structure. This confusion is evident in the images, for example, in Fig. 2, where it is difficult to locate even one 5-coordinated morphological unit.

To solve this problem, Klug and Finch (1965) selected fields in which the stain was relatively thin, and they were able to find images of particles of human papilloma virus which were dominantly stained on one side only. In all cases where two 5-coordinated morphological units could be unequivocally identified, the pattern between them could be seen in terms of steps between morphological units as two out and one to the right (5—6—6$\searrow$5, see Fig. 5). The consistency of this pattern, apart from showing the presence of a definite structure based on a T = 7 lattice, also showed it to be of one particular hand. In order to know which hand of the lattice was present it was necessary to determine which side of the virus particle was being stained. By tilting specimens in the electron microscope it was shown to be the side nearest to the

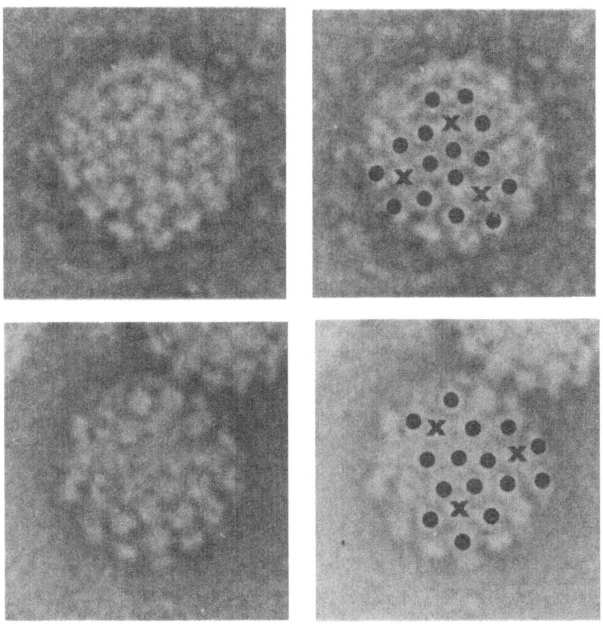

Fig. 5. Electron microscope images of rabbit and human papilloma virus particles, showing the surface units on one side fairly free of superposition effects. On the right-hand copies, the 5-coordinated (X) and 6-coordinated (●) morphological units are marked. By comparison with the models shown in Fig. 4, the rabbit virus has a T = 7l arrangement, the human virus, T = 7d.

carbon film on which the virus particles were sitting. It was thus concluded from the conventions of loading the specimen grids in the microscope and of printing the micrographs that the human papilloma virus is based on the $T = 7d$ icosahedral surface lattice. A similar investigation of the rabbit papilloma virus showed it to be based on the left-hand version of the lattice, $T = 7l$ (Finch and Klug, 1965).

These lattice determinations were based on an extremely small fraction of the papilloma images. The need for dominantly one-side images narrowed the search to fields of thinly stained particles without a complete covering of stain, where virus particles tend to flatten and distort on the support film. Of those that remained reasonably intact, only a few were in a favorable orientation to show two unequivocally 5-coordinated units. However, once the 72-morphological-unit structure was determined, many of the more striking images that occurred in more thickly stained regions were immediately interpretable as two-side images of this structure. Probably the most striking superposition pattern is that shown in Fig. 5; it has three clear, central, large "units" surrounded by triangular terraces of smaller "units." Klug and Finch (1965) recognized this as the two-side image of the 72-morphological-unit structure seen in the direction of a threefold symmetry axis, and its main features were reproduced in a shadow cast by a skeletal model of the structure in light parallel to a threefold axis. However, this shadowgraph representation of an image has several drawbacks. Probably the most serious is that overlapping shadows from the two sides of the model do not appear twice as black, whereas in the two-side electron microscope image the increased contrast of overlap regions is an important factor in the appearance of detail. The model itself must be fairly confined to a thin shell or no light will be able to pass through, and also any extraneous components needed to hold the model together will also cast shadows. Caspar (1965) simulated the image of a negatively stained particle by an analogue method in which a model of the surface structure was coated in plaster of paris (the analogue negative stain) and then X-radiographed. Although quite close agreement was obtained with the electron microscope image, this method is rather tedious. A method of producing superposition patterns computationally from a computed model has been found very useful. Apart from not having to build a physical model for each modification tried, it is possible to compute the projection in any specified direction.

The steps in building a computational model, by comparing its projection with that of the threefold, two-side image are shown in Fig. 6. Briefly, units representing the individual protein subunits are positioned according to the symmetry of the $T = 7$ lattice and the array is

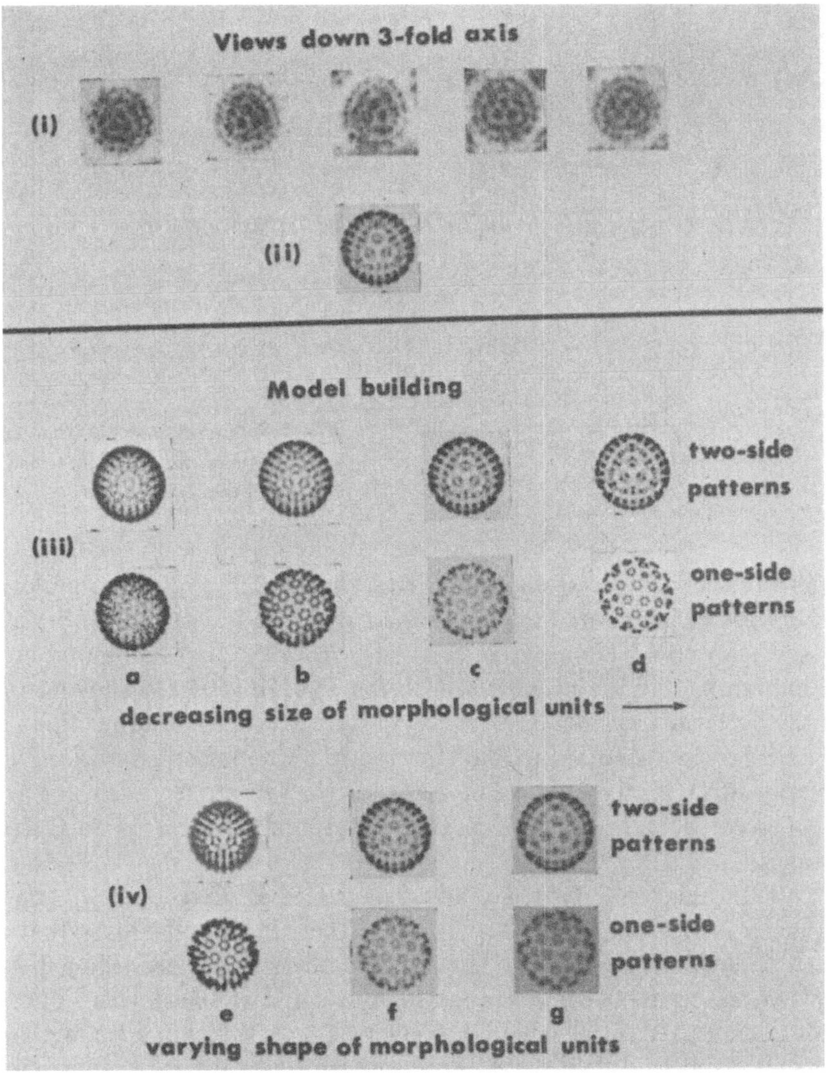

Fig. 6. Simulation of papilloma virus images by computer (Klug and Finch, 1968). (*i*) Examples of the characteristic two-side image of human papilloma virus seen in a direction close to a threefold axis. (*ii*) Simulated image [(*iv*) f] which gives best agreement with *i*. (*iii* and *iv*) Computed patterns for models based on the $T = 7d$ lattice. In *iii* the diameter of the morphological units is varied; in *iv*, their radial extent.

stored in a computer. The projection of this array down any particular direction, in this case down the threefold axis, can then be calculated, displayed on a screen, and photographed. The positions and the radial extents of the structure units in the model can then be altered until the best correspondence between the simulated threefold image and the

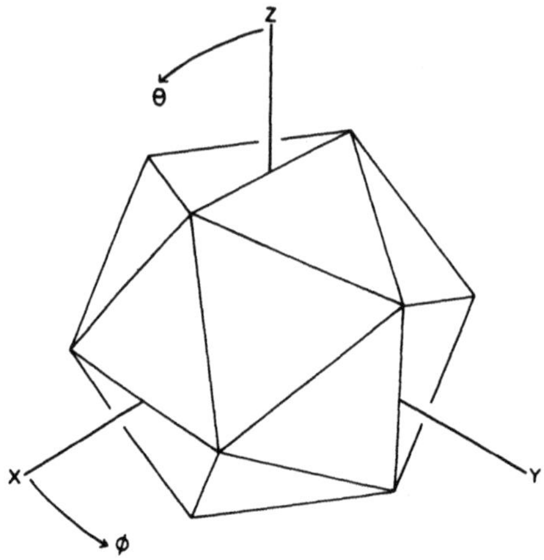

Fig. 7. The relationship between the spherical coordinate frame and the setting of the icosahedron.

electron micrograph is obtained. From the resulting model, a gallery of superposition patterns for comparison with, or identification of, other images can now be computed. A gallery from the best computed model of human papilloma virus, based on the $T = 7d$ lattice is shown in Fig. 8; the patterns are calculated every 3° for a 90° span of projections perpendicular to a twofold axis of the model. This latter limitation yields superposition patterns with left–right mirror symmetry, a feature which is of great help in identifying the corresponding electron microscope images.

With this gallery, Klug and Finch (1968) were able to confirm from two-side images the structure of the papilloma viruses that had been deduced from the very few suitable one-side images. They did this by taking electron micrographs of the same field of virus particles before and after tilting about a known tilt axis through a known angle. Images from this field were chosen which had mirror or approximate mirror symmetry about a line parallel to the tilt axis. Thus superposition patterns could be selected from the gallery which were in closest agreement with the images of the same virus particles before and after tilting (Figs. 8 and 9). It was found that for each particle the angle of rotation to get from one pattern to the other was the same as that used in the tilting experiment and, moreover, that the observed sense of rotation was consistent only with the $T = 7d$ structure for human and the $T = 7l$ structure for the rabbit papilloma viruses.

The method of computing model structures and simulating images by means of superposition patterns has proved very useful in arriving at a reasonably close idea of the surface structure of a virus. However, there are obvious limitations to the degree to which one can represent the locations of individual protein subunits by a series of points or spheres. A more objective way of deducing the morphology of a virus from electron micrographs is now available by means of the technique of three-dimensional image reconstruction which has been applied to

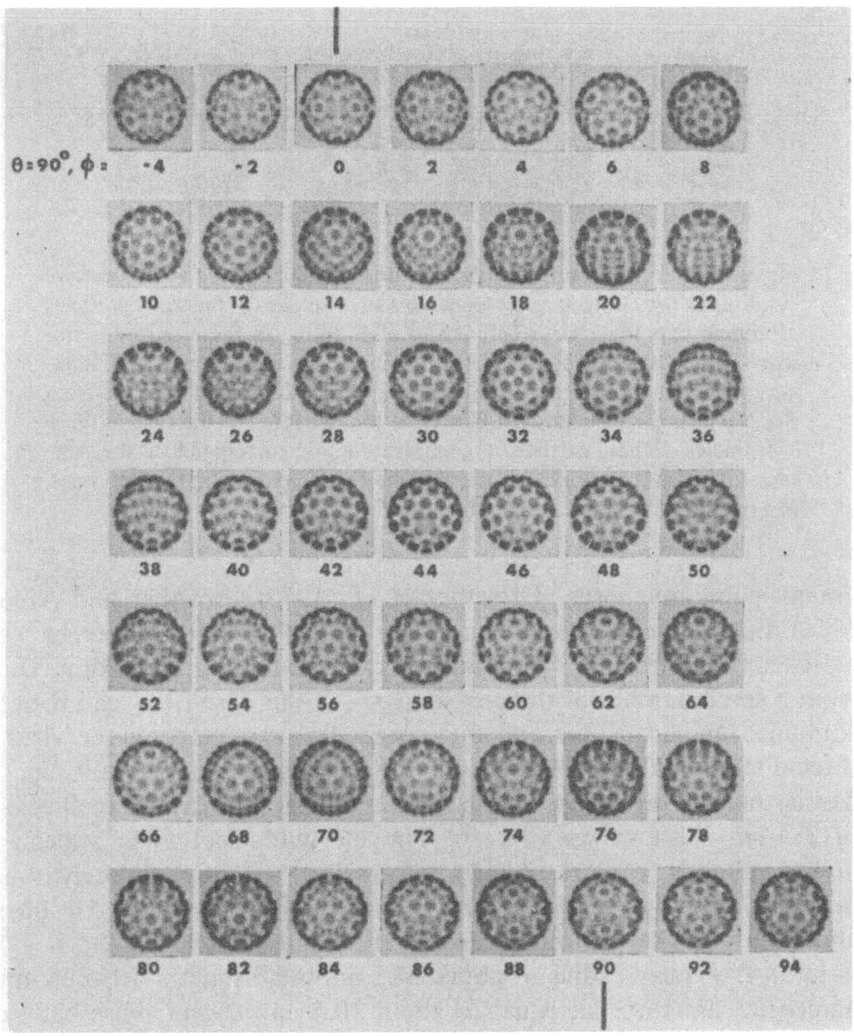

Fig. 8. Gallery of superposition patterns computed for the model g in Fig. 6 for directions of view perpendicular to a twofold axis of the model (i.e., within the plane $\theta = 90°$) (Klug and Finch, 1968). The patterns at $\phi = 0$ and $90°$ correspond to views exactly along the twofold axes of the model, and those at $\phi = 32°$ and $70°$ correspond to views close to the fivefold and threefold axes, respectively; for coordination system see Fig. 7.

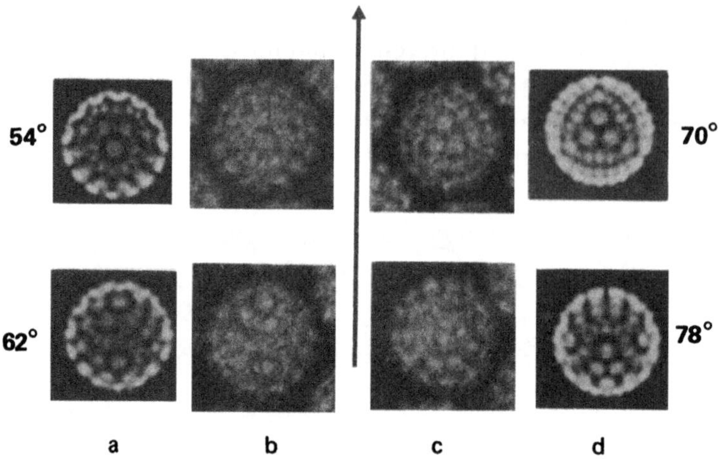

Fig. 9. Determination of the hand of the human papilloma virus structure by tilting in the electron microscope. (b and c) Images of the same particles of human papilloma virus before and after tilting in the electron microscope by about 20°. The tilt is in an anticlockwise sense from b to c, looking in the direction of the arrow. (a and d) The computed patterns from Fig. 8 which correspond most closely to the images. The increase in the ϕ coordinates of these patterns in going from a to d corresponds to the same sense of tilt as that applied to the virus particles, confirming that the virus, like the model structure, is based on the $T = 7d$ lattice.

human papilloma virus (Crowther *et al.*, 1970; Crowther and Amos, 1971). In this method, virus images are tested for their consistency with icosahedral symmetry, a measure of their state of preservation. Data from a small number of the best images—in this case five—can then be combined and from this and the icosahedral symmetry of the virus a three-dimensional reconstruction of the images can be built up. A density map of the reconstruction of human papilloma virus to spacings of 3.7 nm is shown in Fig. 10. The morphological units appear as hollow, slightly conical, peglike protrusions of approximately 9 nm diameter, and they extend from an inner radius of about 21 nm to about 28 nm. They fall fairly closely at the lattice points of the $T = 7d$ lattice. At a mean radius of about 24.5 nm the distances between morphological units are all equal to about 10.4 nm, except those between pairs of hexamers related by a fivefold axis, which are about 11.8 nm. There is no measurable difference between the radial positions of the hexamers and those of the pentamers; thus, the best representation of the surface lattice corresponding to human papilloma virus is the geodesic sphere shown in Fig. 3a, rather than the deltahedron, but the lattice is opened up somewhat around the fivefold axis. In the reconstruction, the pentamers do not appear to be smaller than the hexamers, as one would expect if they were built from the same type of protein subunits; in fact, they are marginally bigger. However, the reliability of this sort of feature is difficult to judge from one reconstruction, and only if it were a consistent feature of many reconstructions from different images could one justify it as evidence for differences in the subunits making up the pentamers and hexamers.

The reliability of the reconstruction is confirmed by comparing projections of it in appropriate directions with the characteristic electron microscope images. A gallery of such comparisons is shown in Fig. 11. There is good agreement between the two sets of images, allowing for the difficulty in choosing an absolute density level in the reconstruction and for the difference in photographic response in the two sets. The reconstruction yields more regular images since it represents the symmetrized average over a number of particles.

2.3.4. Associated Structures

The dominant particle found in the light band after density-gradient centrifugation is an isometric shell of about the same size as the virus particle. Electron microscope images of negatively stained specimens show the same types of characteristic images as the virus

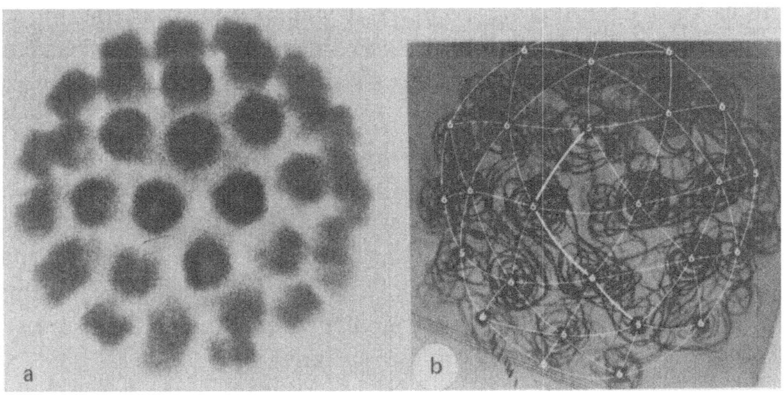

Fig. 10. Three-dimensional reconstructions of human papilloma virus from electron micrographs (Crowther and Amos, 1971). In each case only half of the virus is shown, viewed from approximately the same direction. In (a) data to a Fourier cutoff of 3.7 nm were included; (b) is a contour plot of a lower-resolution (6 nm) reconstruction on which part of the $T = 7d$ lattice has been superposed.

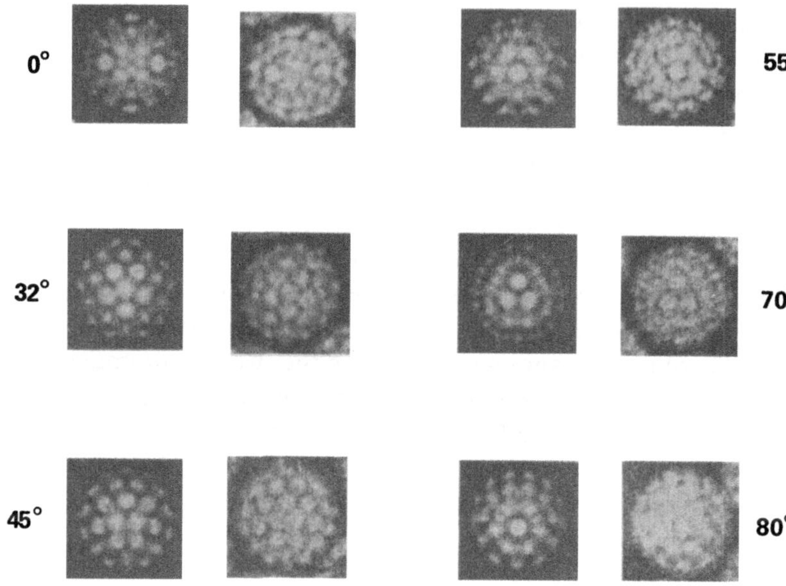

Fig. 11. A comparison of the projections in different directions of the reconstruction of human papilloma virus shown in Fig. 10a (left-hand member of each pair) with images of virus particles in corresponding orientations (right-hand members) (Crowther and Amos, 1971). The values of θ are 90° in each case and the values of ϕ are marked alongside the projections.

particle except that they are completely penetrated by the stain. Their images do appear slightly smaller than those of virus particles, but no discontinuities occur in arrays of virus particles when these empty particles occur; thus, the smaller image is probably a result of the greater stain penetration or perhaps of shrinkage upon drying in the electron microscope. They are apparently empty capsids from which the internal components either have leaked or were omitted during assembly (see Sect. 3.3.3). Occasionally, elongated versions of these shells and particles with a similar substructure but of smaller diameter occur; examples of all these can be seen in Fig. 1, together with some of the more striking structures that occur—the tubular variants.

The hollow, tubular structures associated with the papilloma viruses have been studied in detail by Kiselev and Klug (1969), who used the techniques of optical diffraction and optical filtering to analyze electron microscope images. The tubes fall into two classes, wide ones with diameters of about 50 nm, similar to that of the virus particles, and narrow tubes with diameters of about 30 nm. The optical diffraction patterns from images of wide tubes show approximate local hexagonal symmetry, and the filtered images show distinct hexagonal morphological units arranged in an approximate hexagonal lattice.

The images of narrow tubes give more complex diffraction patterns which show them to be of two related types, a zero-start version which has a threefold rotational symmetry axis and a one-start helical version of the same lattice. The filtered images from these narrow tubes again show distinct morphological units, but in this case they are pentagonal in shape (Fig. 12). The local arrangement of the pentamers is based on the planar tesselation shown in Fig. 13, in which pentagons are packed together as closely as possible, each touching three others. Cylindrical surface lattices based on this tesselation can be formed by rolling up the planar arrangement into a cylinder, preserving the local pattern along the joining line. It was possible to account for both of the surface lattices of the pentamer tubes in this way with little distortion.

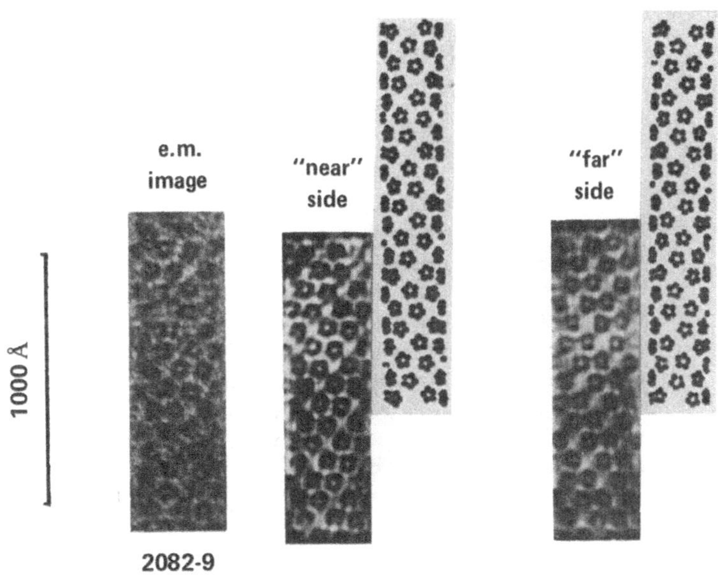

Fig. 12. Electron micrograph and optically filtered one-side images of a narrow tube of the zero-start type from human papilloma virus (Kiselev and Klug, 1969). The arrangement of the morphological units (reproduced black) is the same as that in the computer-generated projections of a model structure based on the pentamer arrangement shown in Fig. 13, which are reproduced adjacent to the one-side images. Although the morphological units in the filtered images are often blurred, they do in places have pentagonal shapes which are in orientations consistent with the model structure.

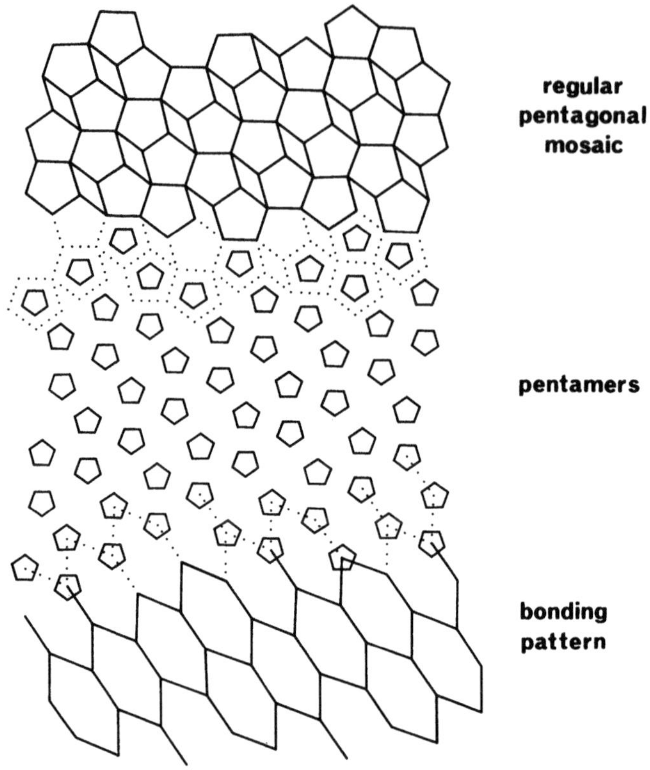

Fig. 13. (Top) A pentagonal tesselation in which each pentagon has three identical neighbors related to each other by incomplete fivefold rotational symmetry. The incomplete symmetry allows the formation of an extended lattice. (Middle) The passage from the above packing of pentagons to an ideal lattice of pentamers representing the observed morphological units in narrow tubes (Fig. 12). This planar lattice can be curved into cylindrical lattices which, with only small distortions, reproduce the observed symmetries in the pentamer tubes.

The two types of pentamer tubes are variants of the same structure, but the coexistence of these with the tubes of hexamers raises the possibility that the two types of morphological unit in the virus particle, the pentamers and the hexamers, are built from different protein-structure units (see Sect. 3.2). The arrangements in the two types of tubes are not, however, inconsistent with the possibility that both are built from the same type of subunit with the same local bonding pattern. On this basis one would expect one type of tube to be thermodynamically the most favorable, although it is possible that once formed, the less stable tube may be difficult to unlock.

Why tubular variants arise is not known. One may postulate that they result from mistakes in the normal capsid assembly—the incorporation of a stray morphological unit, for example, may form a structure which acts as a nucleus for a continuous tubular growth. It is possible that the tubular forms are the naturally favorable aggregates of the hexamers and pentamers in the absence of other material but that normally they aggregate about a core which provides the bonding that makes the normal capsid structure the most energetically favorable.

2.3.5. Discussion of the Structure

From the results described above, the capsids of both human and rabbit papilloma viruses consist of 72 morphological units located at the vertices of a $T = 7$ icosahedral surface lattice. If they are built according to the principles of quasiequivalence described by Caspar and Klug (1962), this surface structure would be built up by the hexamer and pentamer clustering of 420 protein subunits which are either identical or sufficiently similar that they make virtually the same bonding patterns with their neighbors. Some limitations to the extent of the nonidentity of the subunits can be inferred on the basis of symmetry. If the particles possess strict icosahedral symmetry (the consistency of the structure with the $T = 7$ icosahedral surface lattice strongly suggests this and there is no evidence to the contrary), then all the 6-coordinated morphological units are identical and the 5-coordinated morphological units have fivefold symmetry. The 6-coordinated morphological units appear hexagonal and in the wide tubular variants they lie on a hexagonal lattice; thus, it seems reasonable to deduce that they have hexagonal symmetry. On this basis, then, one can conclude that the 60 6-coordinated morphological units are made up of 360 identical protein subunits. The remaining question is whether the 60 protein subunits comprising the 12 pentamers are different. Although the two types of tubular variants suggest that this may be so, one cannot yet make this a firm decision on structural grounds.

3. SV40 AND POLYOMA VIRUS

3.1. Biology of the Viruses*

Viruses of the polyoma group, of which the best known are polyoma virus and simian virus 40 (SV40), have a good deal in common

* See N. P. Salzman and G. Khoury, Chap. 2, Vol. 3 of this series for a more detailed discussion.

with the papilloma viruses. Their virus particles are smaller in size than those of papilloma viruses but they are otherwise very similar in structure. Likewise, the virus DNAs have the same superhelical configuration but are correspondingly smaller.

These viruses are widely distributed in their natural hosts, mice for polyoma and monkeys for SV40. Similar viruses are widely distributed in human populations. In the case of mice polyoma infection occurs early in life and does not seem to cause tumors in the normal life span. The great advantage of polyoma and SV40 as model tumor viruses, in contrast to the papilloma viruses, has been the ability to grow them in tissue culture and to obtain lytic and transforming interactions by a suitable choice of tissue culture cell.

3.2. Composition

3.2.1. DNA Content and Molecular Mass

The particles of polyoma and SV40 contain double-stranded DNA with a molecular mass of 3.6×10^6 daltons. Estimates of the molecular mass have ranged from 2.3 to 3.7×10^6 daltons (Weil and Vinograd, 1963; Crawford, 1964; Caro, 1965; Crawford and Black, 1964; Anderer *et al.*, 1967; Tai *et al.*, 1972). Some of this variation is probably due to the use of different methods and some to the presence of short, defective molecules in the DNA preparations (Yoshiike, 1968; Thorne *et al.*, 1968). The most reliable estimate for the molecular mass of SV40 DNA is 3.6×10^6 daltons (Tai *et al.*, 1972). Polyoma virus DNA is likely to have the same molecular mass. Length measurements of SV40 DNA and polyoma DNA by electron microscopy with the same internal standard of PM2 bacteriophage DNA give values of 3.4×10^6 ($\pm 0.2 \times 10^6$) daltons for both DNAs (Crawford *et al.*, 1975).

The DNA is associated with about an equal amount of histone, as mentioned below, and comprises 12% of the weight of the virus particle. Calculating from these figures the total amount of coat protein per particle would be 21×10^6 daltons. Divided by the figure of 420 subunits, this gives 50,000 daltons for the molecular mass of the subunit, consistent with estimates of polypeptide molecular mass derived from gel electrophoresis.

3.2.2. Polypeptides

Early analyses of polyoma virus protein by gel electrophoresis showed a single major band (Thorne and Warden, 1967). Subsequently, more and more polypeptides, up to a dozen, were detected in varying amounts, in addition to the major capsid protein. Some of these are no doubt contaminants, but it is still not certain which are essential components of the virus and where they fit in the structure. The most generally accepted view of the polypeptide composition is that there are six polypeptides, VP1 to VP6 in order of decreasing size (Table 2).

3.2.2a. VP1

This is the major capsid protein, with a molecular mass of between 42,000 and 50,000 daltons, that of SV40 usually appearing to be slightly smaller than that of polyoma (Girard *et al.*, 1970; Estes *et al.*, 1971; Hirt and Gesteland, 1971; Roblin *et al.*, 1971). The actual values obtained in different laboratories vary slightly, but the variation is no more than that seen with other proteins. This polypeptide comprises about 75% of the virus particle. Amino acid analyses of virus particles are, therefore, quite similar to analyses of isolated VP1 (Murakami *et al.*, 1968; Schlumberger *et al.*, 1968; Kass, 1970; Greenaway and Levine, 1973). The analyses do not show any very unusual features. In addition to amino acids, VP1 contains phosphate, both in SV40 (Tan and Sokol, 1972) and in polyoma (Ponder and Crawford, unpublished results). The amount of phosphate present corresponds to less than one molecule of phosphate per molecule of VP1, e.g., about 0.01 phosphate

TABLE 2

Molecular Weights of SV40 and Polyoma Virus-Particle Polypeptides

	SV40				Polyoma			
	Girard *et al.*, 1970	Barban and Goor, 1971	Estes *et al.*, 1971	Hirt and Gesteland, 1971	Murakami *et al.*, 1968	Murakami *et al.*, 1968	Roblin *et al.*, 1971	Hirt and Gesteland, 1971
VP1	45,000	45,000	43,000	42,000	50,200 ± 4,000	46,000 (calculated from particle molecular weight)	48,000 (P2)	43,000
VP2	35,000		32,000	35,000			35,000 (P3)	
VP3	25,000	29,000	23,000	25,000			23,000 (P4)	
VP4	18,000	16,000	14,000	16,500			19,000 (P5)	
VP5	14,000		12,500				17,000 (P6)	
VP6	10,000		11,000	11,500			15,000 (P7)	

per VP1 in the case of SV40 (Tan, personal communication) and about 0.1 phosphate per VP1 in polyoma. Polyoma VP1 may also contain some carbohydrate since it stains with periodic acid–Schiffs reagent and incorporates small amounts of ^{14}C from ^{14}C-glucosamine (Crawford, 1973). The presence of carbohydrate will affect estimates of molecular mass based on electrophoretic mobility of the polypeptides on gels, making them a little higher than the true value. Very little is known of the amino acid sequence of VP1 beyond the fact that the N-terminal sequence of SV40 VP1 is Ala-Pro-Thr-Lys-Arg-Lys-Gly- and that of polyoma is Ala- (Lazarides *et al.*, 1974). This differs from the previous report that the N terminal of polyoma VP1 is blocked (Murakami *et al.*, 1968), perhaps reflecting differences in the methods used to open up the protein to make the N terminal available for dansylation.

VP1 is clearly virus coded, as would be expected for the major capsid protein. This was shown by the synthesis of VP1 related material in an *in vitro* coupled protein synthesizing system derived from *E. coli* (Crawford and Gesteland, 1972). Purified polyoma virus DNA was added to a system containing *E. coli* RNA polymerase and *E. coli* ribosomes, etc. Most of the polypeptides synthesized were smaller than VP1 but tryptic peptide fingerprints showed that they were VP1 related. Work with other cell-free systems has now shown that a polypeptide with the same electrophoretic mobility as VP1 can be made *in vitro* in response to polyoma cRNA (RNA synthesized *in vitro* with *E. coli* RNA polymerase at high ionic strength). Tryptic peptide fingerprints confirm its identity with authentic VP1 from virus particles (Mangel, Smith, and Bayley, unpublished results). The virus particles of some temperature-sensitive mutants of polyoma have recently been examined by Friedmann and Eckhart (1975). Two members of the main complementation group of the late mutants *ts* 441 and *ts* 610 had alterations in the tryptic peptide fingerprints of virus particle proteins. Although it could be argued that these alterations are distinct from, and in addition to, the lesion which makes the virus particle heat labile the most straightforward explanation would be that the *ts* lesion results in the peptide difference seen. This is again consistent with VP1 being virus coded.

In the past, several reports of other properties for the main capsid protein of SV40 virus have appeared. Initially, two proteins of molecular weights 16,900 and 16,400 were found (Anderer *et al.*, 1967; Schlumberger *et al.*, 1968). More recently, two proteins with masses of about 45,000 daltons, separable by isoelectric focusing, were obtained from SV40 (Barban and Goor, 1971). It seems likely these were different forms of VP1 rather than distinct polypeptides.

3.2.2b. VP2 and VP3

Several polypeptides with masses 20,000–35,000 daltons are usually seen. Two of these, often appearing as doublets on the gels, are fairly constant, but their nature and function is still not clear. In polyoma these two polypeptides appear to be closely related to each other, and quite distinct from VP1 (Fey and Hirt, 1975). This is at variance with a previous report that VP2 and VP3 showed many peptides in common with VP1 when tryptic peptide fingerprints of the various proteins were compared with each other (Friedmann, 1974). It is possible that VP1 in some strains of polyoma is easily broken down and the fragments then run with VP2 and VP3. Amino acid analyses of VP2 and VP3 of SV40 showed that they were distinct from VP1 (Greenaway and Levine, 1973).

It is not clear whether VP2 and VP3 are virus coded or not. Since they form an integral part of the virus particle this might be taken to mean that they must be virus coded. However VP4, 5, and 6 are definitely part of the virus particle but seem not to be virus coded (3.2.2c). Most of the VP2 and VP3 that appears in virus particles seems to be made after virus infection although in some prelabeling experiments VP3 behaved as though it was a mixture of two polypeptides, one made before and one after infection (Frearson and Crawford, 1972).

As yet no *ts* mutants have been characterized with tryptic peptide alterations in VP2 or VP3. Synthesis *in vitro* may provide a clear indication of virus coding but since these are minor components of the virus it is technically more difficult to carry out the necessary comparisons. There are indications that polypeptides with the same mobilities as VP2 and VP3 are synthesized in the wheat germ cell-free system in response to polyoma cRNA (Mangel, Smith, and Bayley, unpublished results).

VP2 and VP3 may have one or both of two functions. Since VP3 appears to be closely associated with the DNA, it may exert some control over its expression (Huang *et al.*, 1972). The nucleoprotein complex isolated from SV40-infected cells has an unwinding protein associated with it (Sen and Levine, 1974), although there is no evidence that it is part of a complex or that it corresponds to VP3. The properties of some *ts* mutants such as *ts* 3 of polyoma and *ts* 101 of SV40 could be explained by association of a protein with the virus DNA and the inability of DNA carrying an altered protein to become active after infection at the nonpermissive temperature. (Robb and Martin, 1972; Francke and Eckhart, 1973). Again there is no evidence to equate this protein definitely with VP3.

The other possible function for VP3 might be to make up the pentons of the virus particle, those subunits surrounded by five neighbors. This is a possibility raised by the particle structure but not required by it, in that VP1 could form both pentons and hexons. The disposition of VP2 and VP3 in the virus particle is clearly different from that of VP1. Treatment of virus particles with SDS in the absence of reducing agents converts full particles to empty shells with the release of the DNA with its associated VP4, 5, and 6. VP2 and 3 are also lost but it is not clear whether this is due to their association with the DNA or to the removal of the pentons from the virus particle (Walter and Deppert, 1975). The main structure of the empty shell is made up of VP1 and until it can be decided whether such shells have or have not lost their pentons this point cannot be resolved. Particles examined immediately after treatment with SDS, but without separation from the released polypeptides, still have the normal appearance of empty particles with pentons rather than holes on the fivefold axes (our own observations).

3.2.2c. VP4, VP5, and VP6

The other polypeptides found in virus particles all have molecular masses of less than 20,000 daltons and seem to correspond to histones of the host cell. This was suggested by their general similarity of properties on SDS and urea gels (Estes *et al.*, 1971; Roblin *et al.*, 1971). Prelabeling studies showed that histones made before infection could later be incorporated into virus particles (Frearson and Crawford, 1972). In addition the comparison of the virus-derived polypeptides with host-cell histones showed that they gave the same methionine-containing tryptic peptides (Frearson and Crawford, 1972), and this has subsequently been extended to all the tryptic peptides (Fey and Hirt, 1975). Whether this group of polypeptides is separated into three or four bands (Lake *et al.*, 1973) depends on the gel system used, but there is general agreement that all the classes of cellular histones, with the exception of F1, are present. The correspondence of the individual histones with the virus-derived basic proteins is probably as follows: VP4 = F3 plus F2b [separated in Lake *et al.* (1973) as two bands], VP5 = F2a2, and VP6 = F2a1. The histones are closely associated with the virus DNA, as they are with the host DNA in chromatin. Complexes of virus DNA with histones and other proteins can be isolated from the infected cell (Green *et al.*, 1971). Alkali disruption of SV40 particles also gives rise to a similar complex (Anderer *et al.*, 1968; Huang *et al.*,

1972). Much of the viral DNA in infected cells is in such complexes, both in polyoma and SV40 (Green *et al.*, 1971; White and Eason, 1971; Goldstein *et al.*, 1973). DNA replication seems to occur in the complexes since replicating DNA appears in more rapidly sedimenting complexes than do completed DNA molecules (Goldstein *et al.*, 1973; Seebeck and Weil, 1974). However, most of the transcription *in vivo* occurs in some other form of viral DNA which does not appear in the Triton supernatant with the bulk of the virus DNA (Shmookler *et al.*, 1974). A small amount of VP1 is also found in association with those deoxyribonucleoprotein complexes, and this could be an indication of their involvement in the assembly of virus particles.

3.3. Structure

3.3.1. Historical Summary

Electron micrographs of negatively stained polyoma virus were first published by Wildy *et al.* (1960). The appearance of the images is very similar to that of the papilloma viruses (Fig. 12), the surface being covered with close-packed, large, morphological units, except that the particle diameter is smaller, about 43 nm (Crawford and Crawford, 1963). Wildy and co-workers recognized both 5- and 6-coordinated morphological units and proposed a model with icosahedral symmetry and consisting of 42 morphological units. No detailed analysis of the images, however, was presented, and this model was questioned by Caspar and Klug (1962) and by Mattern (1962). Mattern *et al.* (1963) considered a 72-unit model for the related K virus but favored a model consisting of 92 morphological units.

To try to resolve between these structures, Howatson and Crawford (1963) counted the capsomers in puddles where capsids presumably had disrupted on the specimen film during the preparation and staining procedure for microscopy. These puddles formed quite discrete collections in which the capsomers were often spread out and easily counted, and Howatson and Crawford arrived at a mean count close to 42 capsomers per puddle. However, this was in turn criticized by Klug and Finch (1965), who questioned the relevance of the mean count. Since regular capsids would be composed of a constant number of capsomers, if all the capsomers per puddle originated from one capsid, then the variation in numbers per puddle could only mean that not all the capsomers were trapped in the puddle upon disruption. On this basis the maximum number of capsomers per puddle found would be a closer approach to the number per capsid.

3.3.2. Morphology

Following the establishment of the $T = 7$ structure of the papilloma viruses, Klug (1965) showed that the earlier-published electron micrographs of polyoma virus [from Wildy *et al.* (1960) and Crawford *et al.* (1962)] which showed clearest surface detail (i.e., those arising from particles stained dominantly on one side of the particle) were also consistent with a $T = 7$, 72-morphological-unit structure and inconsistent with the earlier-proposed 42-unit model. In many cases Klug was able to identify two 5-coordinated morphological units and show that the path between them was of the form 5—6—6⟍5 (see Fig. 4). Similar paths were recognized in one-side images of SV40 by Anderer *et al.* (1967), who thus showed that this virus was based on the $T = 7d$ lattice.

Finch (1974) has obtained images of negatively stained polyoma virus over holes in the supporting film (Fig. 14). In these fields the particles are completely enveloped in stain, and thus the resultant images are true two-side images arising from the superposition of detail from both near and far sides of the virus particles. Many of these images are recognizable from the gallery of superposition patterns

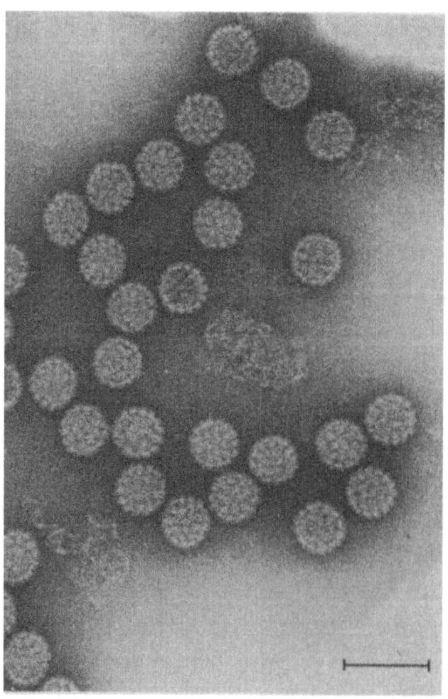

Fig. 14. Electron micrograph of polyoma virus particles suspended in stain over a hole in the carbon substrate. This method ensures that the images are superposition patterns of detail equally from the near and far sides of the particles and thus are suitable for use in three-dimensional reconstruction. The film of stain was broken and contracted and, although the particles are somewhat compressed, the distortion in this field is fairly isometric (bar = 50 nm).

32°

44°

62°

70°

82°

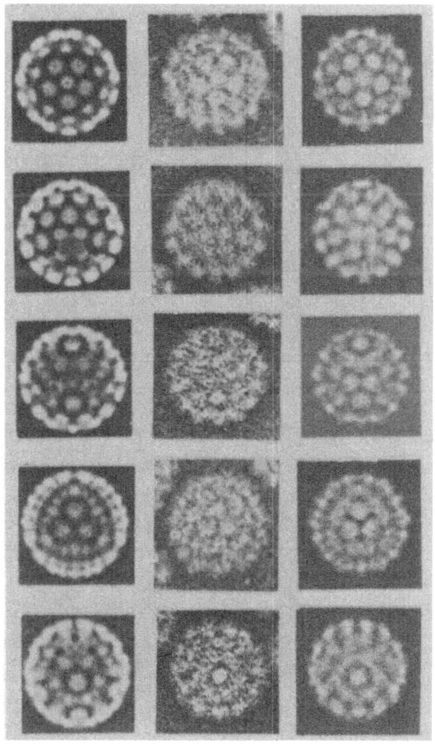

a b c

Fig. 15. (a) Corresponding computed patterns from the gallery in Fig. 8. (b) Characteristic two-side images of negatively stained polyoma virus particles. (c) Projections of the three-dimensional reconstruction of polyoma virus shown in Fig. 16. In each case $\theta = 90°$ and the values of ϕ are given (Finch, 1974).

shown in Fig. 8, computed on the basis of hexamer–pentamer clustering of structure units on the $T = 7$ lattice for comparison with the images of papilloma virus (Fig. 15), confirming that polyoma has a structure of this type. Tilting experiments with polyoma virus similar to those described for the papilloma viruses have shown that polyoma is based on the $T = 7d$ lattice.

The three-dimensional image-reconstruction technique has also been applied to electron micrographs of polyoma virus (Finch, 1974). A view of the density plot of the reconstruction is shown in Fig. 16. The 72 morphological units over the capsid surface all lie at about the same radius and are centered at the lattice points of the $T = 7d$ icosahedral surface lattice. At a mean radius of contrast of 19 nm, these distances

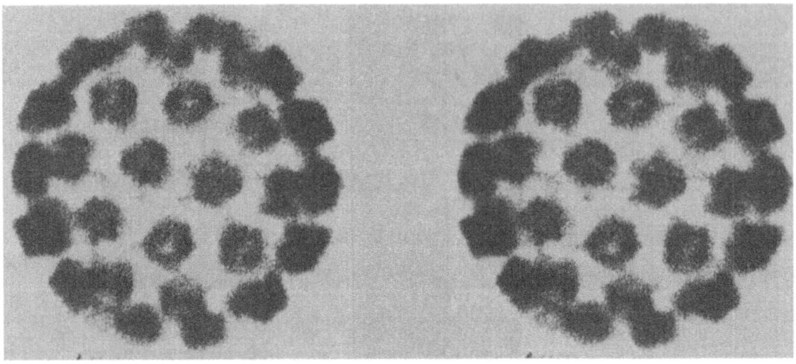

Fig. 16. Stereoview of a reconstruction of polyoma virus, including data to a Fourier cutoff of 2.5 nm (Finch, 1974). The view is close to a twofold symmetry axis and only the top half of the reconstruction is shown. Two 5-coordinated morphological units can be seen clear of the periphery, toward the upper right and lower left, and the path between them is two steps out and one to the right, corresponding to the icosahedral $T = 7d$ lattice.

between the morphological units are all about 7.5–8.0 nm, except for the distances between the hexamers which immediately neighbor pentamers, which are about 9.3 nm apart. Thus, the surface lattice is opened up around the fivefold vertices in a similar fashion to that of the human papilloma virus (Sect. 2.3.3). The morphological units in the polyoma reconstruction are hollow and slightly conical in shape, with diameters of about 5.0 nm, although diameter size is very sensitive to the choice of the contour level chosen as the boundary between protein and stain. The units are first distinguished at a radius of about 18 nm and extend to about 21 nm. The pentamers appear very slightly smaller than the hexamers, but only if this were a consistent feature of many reconstructions could we regard it as structurally significant. The reconstruction shows little indication of regular substructure within the morphological units.

3.3.3. Associated Structures

As with the papilloma viruses, the commonest particle found associated with SV40 and polyoma virus is the empty capsid. Some information on the significance of the empty capsids has been obtained in the case of SV40 in an investigation of the synthesis and assembly of this virus by Ozer (1972) and Ozer and Tegtmeyer (1972). After a short pulse exposure to ^{3}H-lysine during infection, the empty capsids were found to be preferentially labeled, suggesting that these were synthesized independently of virus particles. This was confirmed by the observation that inhibition of DNA synthesis did not affect production of empty capsids although synthesis of complete virus was lowered. In the subsequent chase period, however, radioactive capsid protein was found preferentially in intact virus particles, suggesting that empty capsids were precursors to the virus particles. The presence of internal virus protein associated with empty capsids, however, indicates that at least some result from the breakdown of completed virus particles.

In addition to the empty capsid, Anderer *et al.* (1967) observed three smaller types of isometric shell which they suggested were the $T = 1$, 3, and 4 variants of the empty capsid. Mattern *et al.* (1967) reported similar smaller particles associated with polyoma virus and suggested that they existed inside the complete particle; this seems unlikely from the chemical make-up of the virus, and physically it would leave little room for the DNA–histone core.

Tubular structures similar to those found with papilloma viruses are also associated with SV40 and polyoma. For polyoma, Kiselev and Klug (1969) again found two main classes: wide tubes about 60 nm in diameter built from hexamers and narrow tubes about 30 nm in diameter built from pentamers. Mattern *et al.* (1967) reported three main classes of diameter, with some conversion from one type to another within the same tube.

3.4. Related Human Viruses

3.4.1. Properties of the Viruses

Recently, several viruses have been isolated from human tissues, often from patients who have been under immunosuppression for organ transplants. These viruses are clearly related to SV40 and generally are rather similar in their properties. Three of these viruses were isolated from cases of progressive multifocal leucoencephalopathy (PML), a rare demyelinating disease (Padgett *et al.*, 1971; Weiner *et al.*, 1972). Another virus, BK, was isolated from a renal transplant patient by Gardner *et al.*, (1971). The electron microscopic appearance of these particles is very similar to that of SV40, and BK virus also shows some antigenic cross reaction with SV40 (Howatson *et al.*, 1964; Gardner *et al.*, 1971; Padgett *et al.*, 1971; Penney and Narayan, 1973; Takemoto and Mullarkey, 1973). BK and DAR [case 2 of Weiner *et al.* (1972)] differ from SV40 in that they hemagglutinate (Gardner *et al.*, 1971;

TABLE 3

Molecular Weights of Polypeptides of SV40-Related Human Viruses[a]

		SV40	DAR	BK
VP1	(Vp1)	48,000	48,000	45,000
	(Vp2)	40,000	40,500	—
VP2	(Vp3)	38,000	38,000	36,500
	(Vp4)	32,000	32,000	32,000
VP3	(Vp5)	27,600	28,000	28,000
VP4	(Vp6)	18,500	19,000	18,000
VP6	(Vp7)	15,200	15,000	15,000

[a] The values in this table are from Mullarkey *et al.* (1974) and their numbering is given in parentheses. The numbering of VP1–VP6 indicates the probable correspondence to the numbering used in Table 1 (Estes *et al.*, 1971). Here the molecular weight of SV40 VP1 is taken as 48,000 rather than the value of 43,000 obtained by Estes *et al.* (1971) (Table 2).

Padgett *et al.*, 1971). This is really their only similarity to polyoma, and all other studies show them to be closer to SV40 than to polyoma. In particular, the DNA of DAR virus is similar to that of SV40 in structure and molecular weight and gives a similar pattern of DNA fragments when digested by restriction enzymes from *Hemophilus influenzae* (Sack *et al.*, 1973). Two of the fragments do appear to be different. Probably the other human viruses are less closely related to SV40. This would be consistent with their antigenic relationships, BK being less similar to SV40 than DAR (Mullarkey *et al.*, 1974).

3.4.2. Polypeptide Composition

The overall composition of the virus particles is very similar to that of SV40, as shown in Table 3. A polypeptide of molecular weight 40,500 observed in analyses of SV40 and DAR was referred to as VP2. This polypeptide does not appear in analyses of BK virus, and, as it is about the same size as a prominent polypeptide seen in gels of uninfected-cell extracts, it is possible that it is a cellular polypeptide. The polypeptides of DAR have molecular weights very close to those of SV40, but again BK is less similar, the molecular weight of VP1 being about 45,000 as compared to 48,000 for SV40 (Mullarkey *et al.*, 1974).

ACKNOWLEDGMENTS

It is a pleasure to express our thanks to Mrs. Cilla Conway for her help with preparation of the manuscript.

4. REFERENCES

Anderer, F. A., Schlumberger, H. D., Koch, M. A., Frank, H., and Eggers, H. J., 1967, Structure of simian virus 40. II. Symmetry and components of the virus particle, *Virology* **32,** 511.

Anderer, F. A., Koch, M. A., and Schlumberger, H. D., 1968, Structure of simian virus 40. III. Alkaline degradation of the virus particle, *Virology* **34,** 452.

Barban, S., and Goor, R., 1971, Structural proteins of SV40, *J. Virol.* **7,** 198.

Breedis, C., Berwick, L., and Anderson, T. F., 1962, Fractionation of Shope papilloma virus in cesium chloride density gradients, *Virology* **17,** 84.

Caro, L. G., 1965, The molecular weight of lambda DNA, *Virology* **25,** 226.

Caspar, D. L. D., 1965, An analogue for negative staining, *J. Mol. Biol.* **15,** 365.

Caspar, D. L. D., and Klug, A., 1962, Physical principles in the construction of viruses, *Cold Spring Harbor Symp. Quant. Biol.* **27,** 1.

Crawford, L. V., 1964, A study of Shope papilloma virus DNA, *J. Mol. Biol.* **8,** 489.

Crawford, L. V., 1965, A study of human papilloma virus DNA, *J. Mol. Biol.* **13,** 362.

Crawford, L. V., 1973, Proteins of polyoma virus and SV40, *Brit. Med. Bull.* **29,** 253.

Crawford, L. V., and Black, P. H., 1964, The nucleic acid of simian virus 40, *Virology* **24,** 388.

Crawford, L. V., and Crawford, E. M., 1963, A comparative study of polyoma and papilloma viruses, *Virology* **21,** 258.

Crawford, L. V., Crawford, E. M., and Watson, D. H., 1962, The physical characteristics of polyoma virus. I. Two types of particle, *Virology* **18,** 170.

Crawford, L. V., and Gesteland, R. F., 1973, Synthesis of polyoma proteins *in vitro*, *J. Mol. Biol.* **74,** 627.

Crawford, L. V., Robbins, A. K., Nicklin, P. M., and Osborne, K., 1975, Polyoma DNA replication: location of the origin of different virus strains, *Cold Spring Harbor Symp. Quant. Biol.* **39,** 219.

Crowther, R. A., and Amos, L. A., 1971, Three-dimensional image reconstruction of some small spherical viruses, *Cold Spring Harbor Symp. Quant. Biol.* **36,** 489.

Crowther, R. A., De Rosier, D. J., and Klug, A., 1970, The reconstruction of an object from its projections and its application to electron microscopy, *Proc. R. Soc. Lond. Ser. A. Math Phys. Sci.* **317,** 319.

Estes, M. K., Huang, E., and Pagano, J., 1971, Structural polypeptides of SV40, *J. Virol.* **7,** 635.

Fey, G., and Hirt, B., 1975, Two dimensional fingerprints of tryptic peptides from polyoma virion proteins and mouse histones, *Cold Spring Harbor Symp. Quant. Biol.* **39,** 235.

Finch, J. T., 1974, The surface structure of polyoma virus, *J. Gen. Virol.*, **24,** 359.

Finch, J. T., and Klug, A., 1965, Structure of viruses of the papilloma-polyoma type: III. Structure of the rabbit papilloma virus, *J. Mol. Biol.* **13,** 1.

Francke, E., and Eckhart, W., 1973, Polyoma gene function required for viral DNA synthesis, *Virology* **55,** 127.

Frearson, P. M., and Crawford, L. V., 1972, Polyoma virus basic proteins, *J. Gen. Virol.* **14,** 141.

Friedmann, T., 1974, Genetic economy of polyoma virus: Capsid proteins are cleavage products of same viral gene, *Proc. Natl. Acad. Sci. USA* **71,** 257.

Friedmann, T., and Eckhart, W., 1975, Mutant virion proteins in polyoma temperature sensitive mutants, *Cold Spring Harbor Symp. Quant. Biol.* **39,** 243.

Gardner, S. D., Field, D. M., Coleman, D. V., and Hulme, B., 1971, New human papovavirus (BK) isolated from urine after renal transplantation, *Lancet* **I,** 1253.

Gibson, W., 1974, Polyoma virus proteins: a description of the structural proteins of the virion based on polyacrylamide gel electrophoresis and peptide analysis, *Virology* **62,** 319.

Girard, M., Marty, L., and Suarez, F., 1970, Capsid proteins of simian virus 40, *Biochem. Biophys. Res. Commun.* **40,** 97.

Goldstein, D. A., Hall, M. R., and Meinke, W., 1973, Properties of nucleoprotein complexes containing replicating polyoma DNA, *J. Virol.* **12,** 887.

Green, M., Miller, H., and Hendler, S., 1971, Isolation of a polyoma nucleoprotein complex from infected mouse cell cultures, *Proc. Natl. Acad. Sci. USA* **68,** 1032.

Greenaway, P. J., and Levine, D., 1973, Amino acid compositions of simian virus 40 structural proteins, *Biochem. Biophys. Res. Commun.* **52,** 1221.

Hirt, B., and Gesteland, R. F., 1971, Characterization of SV40 and polyoma virus, *in* "Lepetit Colloquia on Biology and Medicine," Vol. 2, "The Biology of Oncogenic Viruses" (L. Silvestri, ed.), pp. 98–103, North-Holland, Amsterdam.

Howatson, A. F., and Crawford, L. V., 1963, Direct counting of the capsomeres in polyoma and papilloma viruses, *Virology* **21,** 1.

Howatson, A. F., Nogai, M., and ZuRhein, G. M., 1964, Polyoma-like virions in human demyelinating brain disease, *Can. Med. Assoc. J.* **93,** 379.

Huang, E.-S., Estes, M. K., and Pagano, J. S., 1972, Structure and function of the polypeptides in simian virus 40. I. Existence of subviral deoxynucleoprotein complexes, *J. Virol.* **9,** 923.

Kass, S. J., 1970, Chemical studies on polyoma and Shope papilloma viruses, *J. Virol.* **5,** 381.

Kass, S. J., and Knight, C. A., 1965, Purification and chemical analysis of Shope papilloma virus, *Virology* **27,** 273.

Kiselev, N. A., and Klug, A., 1969, The structure of viruses of the papilloma–polyoma type. V. Tubular variants built of pentamers, *J. Mol. Biol.* **40,** 155.

Kleinschmidt, A. K., Kass, S. J., Williams, R. C., and Knight, C. A., 1965, Cyclic DNA of Shope papilloma virus, *J. Mol. Biol.* **13,** 749.

Klug, A., 1965, Structure of viruses of the papilloma-polyoma type. II. Comments on other work, *J. Mol. Biol.* **11,** 424.

Klug, A., and Finch, J. T., 1965, Structure of viruses of the papilloma–polyoma type. I. Human wart virus, *J. Mol. Biol.* **11,** 403.

Klug, A., and Finch, J. T., 1968, Structure of viruses of the papilloma–polyoma type. IV. Analysis of tilting experiments in the electron microscope, *J. Mol. Biol.* **31,** 1.

Lake, R. S., Barban, S., and Salzman, N. P., 1973, Resolutions and identification of the core deoxyribonucleoproteins of simian virus 40, *Biochem. Biophys. Res. Commun.* **54,** 640.

Lazarides, E., Files, J. G., and Weber, K., 1974, Simian virus 40 structural proteins: amino-terminal sequence of the major capsid protein, *Virology* **60**, 584.
Le Bouvier, G. L., Sussmann, M., and Crawford, L. V., 1966, Antigenic diversity of mammalian papilloma viruses, *J. Gen. Microbiol.* **45**, 497.
Mattern, C. F. T., 1962, Polyoma and papilloma viruses: do they have 42 or 92 subunits? *Science* (Wash., D.C.) **137**, 612.
Mattern, C. F. T., Allison, C. A., and Rowe, W. P., 1963, Structure and composition of K-virus and its relation to the "papovavirus" group, *Virology* **20**, 413.
Mattern, C. F. T., Takemoto, K. K., and DeLeva, A. M., 1967, Electron microscopic observations on multiple polyoma virus-related particles, *Virology* **32**, 378.
Melnick, J. L., Allison, A. C., Butel, J. S., Eckhart, W., Eddy, B. E., Kit, S., Levine, A. J., Miles, J. A. R., Pagano, J. S., Sachs, L., and Vonka, V., 1974, Papovaviridae, *Intervirology* **3**, 106.
Mullarkey, M. F., Hruska, J. F., and Takemoto, K. K., 1974, Comparison of two human papovaviruses with simian virus 40 by structural protein and antigenic analysis, *J. Virol.* **13**, 1014.
Murakami, W. T., Fine, R., Harrington, M. R., and Ben Sassan, Z., 1968, Properties and amino acid composition of polyoma virus purified by zonal ultracentrifugation, *J. Mol. Biol.* **36**, 153.
Ozer, H. L., 1972, Synthesis and assembly of simian virus 40. I. Differential synthesis of intact virions and empty shells, *J. Virol.* **9**, 41.
Ozer, H. L., and Tegtmeyer, P., 1972, Synthesis and assembly of simian virus 40. II. Synthesis of the major capsid protein and its incorporation into viral particles, *J. Virol.* **9**, 52.
Padgett, B. L., Walker, D. L., ZuRhein, G. M., Echroade, R. J., and Dessel, B. D., 1971, Cultivation of papova-like virus from human brain with progressive multifocal leukoencephalopathy, *Lancet* **I**, 1257.
Pass, F., and Maizel, J. V., 1973, Wart-associated antigens. II. Human immunity to viral structural proteins, *J. Invest. Derm.* **60**, 307.
Penney, J. B., and Narayan, O., 1973, Studies of the antigenic relationships of the new human papovaviruses by electron microscopy agglutination, *Infect. Immun.* **8**, 299.
Robb, J. A., and Martin, R. G., 1972, Genetic analysis of simian virus 40. III. Characterization of a temperature-sensitive mutant blocked at an early stage of productive infection in monkey cells, *J. Virol.* **9**, 956.
Roblin, R., Härle, E., and Dulbecco, R., 1971, Polyoma virus proteins. I. Multiple virion components, *Virology* **45**, 555.
Sack, G. H. Jr., Narayan, O., Danna, K. J., Weiner, L. P., and Nathans, D., 1973, The nucleic acid of an SV40-like virus isolated from a patient with progressive multifocal leukoencephalopathy, *Virology* **51**, 345.
Schlumberger, H. D., Anderer, F. A., and Koch, M. A., 1968, Structure of the simian virus 40. IV. The polypeptide chains of the virus particle, *Virology* **36**, 42.
Seebeck, T., and Weil, R., 1974, Polyoma virus DNA replicated as a nucleoprotein complex in close association with the host cell chromatin, *J. Virol.* **13**, 567.
Sen, A., and Levine, A. J., 1974, SV40 nucleoprotein complex activity unwinds superhelical turns in SV40 DNA, *Nature* (*Lond.*) **249**, 343.
Shmookler, R. J., Buss, J., and Green, M. H., 1974, Properties of the polyoma virus transcription complex obtained from mouse nuclei, *Virology* **57**, 122.
Spira, G., Estes, M. K., Dreesman, G. R., Butel, J. S., and Rawls, W. E., 1974,

Papovavirus structural polypeptides: comparison of human and rabbit papilloma viruses with simian virus 40, *Intervirology* **3**, 220.

Tai, H. T., Smith, C. A., Sharp, P. A., and Vinograd, J., 1972, Sequence heterogeneity in closed simian virus 40 deoxyribonucleic acid, *J. Virol.* **9**, 317.

Takemoto, K. K., and Mullarkey, M. F., 1973, Human papovavirus, BK strain: Biological studies including antigenic relationship to simian virus 40, *J. Virol.* **12**, 625.

Tan, K. B., and Sokol, F., 1972, Structural proteins of simian virus 40: Phosphoproteins, *J. Virol.* **10**, 985.

Thorne, H. V., and Warden, D., 1967, Electrophoretic evidence for a single protein component in the capsid of polyoma virus, *J. Gen. Virol.* **1**, 135.

Thorne, H. V., Evans, J., and Warden, D., 1968, Detection of biologically defective molecules in component I of polyoma virus DNA, *Nature* (*Lond.*) **219**, 728.

Walter, G., and Deppert, W., 1975, Intermolecular disulfide bonds: an important structural feature of the polyoma virus capsid, *Cold Spring Harbor Symp. Quant. Biol.* **39**, 255.

Watson, J. D., and Littlefield, J. W., 1960, Some properties of DNA from Shope papilloma virus, *J. Mol. Biol.* **2**, 161.

Weil, R., and Vinograd, J., 1963, The cyclic helix and cyclic coil forms of polyoma virus DNA, *Proc. Natl. Acad. Sci. USA* **50**, 730.

Weiner, L. P., Herndon, R. M., Narayan, O., Johnson, R., Shah, K., Rubinstein, L. J., Preziosi, T. L., and Conley, F. K., 1972, Isolation of virus related to SV40 from patients with progressive multifocal leukoencephalopathy, *New Engl. J. Med.* **186**, 385.

White, M., and Eason, R., 1971, Nucleoprotein complexes in SV40-infected cells, *J. Virol.* **8**, 363.

Wildy, P., Stoker, M. G. P., Macpherson, I. A., and Horne, R. W., 1960, The fine structure of polyoma virus, *Virology* **11**, 444.

Williams, M. G., Howatson, A. F., and Almeida, J. D., 1961, Morphological characterization of the virus of the human common wart (*Verruca vulgaris*), *Nature* (*Lond.*) **189**, 895.

Williams, R. C., 1953, The shapes and sizes of purified viruses as determined by electron microscopy, *Cold Spring Harbor Symp. Quant. Biol.* **18**, 185.

Williams, R. C., Kass, S. J., and Knight, C. A., 1960, Structure of Shope papilloma virus particles, *Virology* **12**, 48.

Yoshiike, K., 1968, Studies on DNA from low-density particles of SV40. I. Heterogeneous defective virions produced by successive undiluted passages, *Virology* **34**, 391.

CHAPTER 3

Pseudovirions in Animals, Plants, and Bacteria

H. Vasken Aposhian
Department of Cell and Developmental Biology
College of Liberal Arts
University of Arizona
Tucson, Arizona, 85721

1. INTRODUCTION

1.1. Definitions

The term "pseudovirus" was used first by Michel *et al.* (1967) to describe a particle which is produced during the infection of cultured mouse cells by polyoma virus. This polyoma-related particle was found to contain fragments of mouse DNA encapsidated within the protein coat of polyoma virus. Since the discovery of polyoma pseudovirions by Michel *et al.* (1967) and independently by Winocour (1967*a,b*), pseudovirions have been discovered in preparations of simian virus 40 (SV40) and tobacco mosaic virus (TMV). Although pseudovirions have been found only recently in animal cells and plant cells, they appear to be analogous in many ways to particles containing fragments of host-cell DNA, but no detectable virus-specific DNA, that are produced during the infection of bacterial cells by some bacterial viruses. Although these phagelike particles have not been called pseudovirions in the past, they fit the definition of pseudovirions and will be referred to as such in this chapter. Many of these bacterial pseudovirions have been shown to be capable of generalized transduction. However, generalized transduction has not been demonstrated, as yet, with any pseudovirions of animal- or plant-cell origin.

Pseudovirions are defined as viruslike particles that consist of fragments of host-cell nucleic acid encapsidated within the protein coats of known infectious viruses. The pseudovirions do not appear to contain any of the nucleic acid of the infectious virus to which they are related.

Another type of particle, found in lysates of some bacterial cells, is similar in many ways to pseudovirions and will be called *orphan pseudovirions* for the purposes of this chapter. Orphan pseudovirions similarly contain fragments of only host-cell nucleic acid. The nucleic acid is encapsidated within a protein structure that appears under the electron microscope to be morphologically similar to the structures of other known bacterial viruses. However, no one has been able to demonstrate that these protein structures or coats are gene products of any known infectious virus. The identity of the original virus, the genome of which is responsible for the protein structures of an orphan pseudovirion, is not known. It has not been ruled out that some host changes, independent of any viral genome, might have resulted in the excision of DNA and the synthesis of new host-coded proteins.

The reader is cautioned not to confuse pseudovirions with pseudotype virus. A pseudotype virus has the genome of a known virus encapsidated with the protein coat of another but different type of virus. Pseudotype viruses will not be included in the present chapter.

1.2. Scope of the Chapter

In this chapter the pseudovirions found in animal cells, plant cells, and bacterial cells will be discussed. The information is based on selected papers published up to May 1974 and any preprints that have been made available to me. In certain cases, it will be advisable to discuss pseudovirions in context with the synthesis of the virus to which the pseudovirus coat is related. Therefore, where appropriate, I will include a brief description of the pertinent biology of the virus to which the particular pseudovirus is related. This description will be purposely brief since the parental viruses have been thoroughly reviewed in other chapters in this series. The reader is referred to them for more detailed information about the parental virus.

For each type of pseudovirus, the method of preparation will be reviewed. This will be followed by the evidence for the identification and the origin of pseudoviral DNA and protein coats. Next, the intracellular events related to pseudovirus production and the influence of the type of host cell will be presented. Finally, the biological properties and fate of pseudovirions, if known, will be discussed. Such material will, of course, be a review and evaluation of published information. The last section of this chapter will be speculative and concern the possible roles of pseudovirions in nature.

2. ANIMAL CELL PSEUDOVIRIONS

2.1. Polyoma Pseudovirions

2.1.1. Brief Summary of Polyoma Virus Biology

This section will review some of the biology of polyoma virus that is pertinent to polyoma pseudovirions. This description will be brief and incomplete since it is not the primary subject of this chapter and is the subject of another chapter in this series (see Vol. 3, Chapt. 2, by N. P. Salzman and G. Khoury). The reader is also referred to the excellent Cold Spring Harbor Laboratory monograph entitled "The Molecular Biology of Tumour Viruses" (Tooze, 1973).

Polyoma virus has been extensively studied in recent years as a model system for viral carcinogenesis (Dulbecco, 1969; Tooze, 1973). The virus will induce tumors when injected into newborn mice or hamsters (Stewart *et al.*, 1957; Eddy *et al.*, 1958). At least two types of responses may result when the virus is used to infect cells growing in culture. The infection of mouse cells results predominantly in a lytic (productive) infection. In this type of response, viral progeny are produced, cytopathic effects are observed, and eventually cell death results. A number of experiments have suggested that the polyoma genome may be integrated and excised during the productive type of infection (Babiuk and Hudson, 1972; Ralph and Colter, 1972). Whether integration and excision are obligatory events in polyoma virus reproduction (in productive infections) is far from clear. However, an excision mechanism, whatever it is, is probably related to pseudovirus production.

The second type of response to polyoma infection, called transformation, is seen most commonly when hamster cells are infected with the virus. In this case, neither virus production nor cytopathic effects are observed. However, the clonal morphology of the transformed cells is different. Whereas the uninfected cells are arranged in a well-ordered monolayer, the transformed cells have a disoriented clonal morphology in which there is an increase in saturation density of the cells. In these transformed cells, a small number of viral genomes may be detected using DNA–RNA molecular hybridization techniques (Westphal and Dulbecco, 1968). The exact physical state of the viral genomes in transformed cells is, as yet, unknown.

Three types of DNA have been commonly found in polyoma particles. Polyoma type I DNA has a circular, double-stranded, supercoiled configuration (Dulbecco and Vogt, 1963; Weil and Vinograd, 1963) which sediments at 20 S or 53 S in neutral or alkaline sucrose gradients, respectively. If a single nick is made in one strand of type I DNA, the molecule assumes a relaxed topology and has been designated polyoma type II DNA (Vinograd *et al.*, 1965). This DNA sediments at 16 S in neutral sucrose and at 18 S (single-stranded circles) and 16 S (single-stranded linear molecules) in alkaline sucrose gradients. Double-stranded, linear DNA molecules have been isolated from other polyoma particles and designated polyoma type III DNA. Originally, these type III DNA molecules were believed to represent sequences of polyoma-specific DNA. However, it is now believed that type III DNA consists of fragments of cellular DNA and not polyoma-specific DNA. These fragments are now called pseudoviral DNA and sediment at 14 S in neutral gradients and at 16 S in alkaline gradients. All three types of DNA have a molecular weight of approximately 3×10^6.

The infection of mouse cells by polyoma virus may produce a number of polyoma-related particles: infectious polyoma virions, defective polyoma virions, polyoma pseudovirions, empty polyoma capsids, and abnormal virus particles. The *infectious polyoma virions* usually contain the complete polyoma genome, consisting of supercoiled, double-stranded DNA with a molecular weight of approximately 3×10^6. The *defective polyoma virions* are mutants with large deletions in their DNA (Blackstein *et al.*, 1969; Fried, 1974). When the DNA is extracted from these defective virions, it is found to be double-stranded, circular, and supercoiled. However, the molecular weights of some of these DNA molecules are 50%, while others are 75%, that of polyoma type I DNA. Recently, *deletion–substitution particles* that contain sequences of host DNA covalently attached to polyoma DNA have been reported (Lavi and Winocour, 1974). These defective particles appear to be produced when the polyoma virus used for infection has undergone serial passages in the laboratory at a high multiplicity of infection.

Two laboratories, almost simultaneously, reported the production of *polyoma pseudovirions* in baby mouse kidney (BMK) cells infected with polyoma virus (Michel *et al.*, 1967; Winocour, 1967*a,b*). The pseudovirions contain linear, double-stranded fragments of mouse DNA encapsidated within polyoma coats. The DNA from pseudovirions derived from infected BMK cells has a molecular weight estimated to be between 1×10^6 and 3×10^6 (Michel *et al.*, 1967). The pseudovirions from infected primary mouse embryo (PME) cells appear to have a DNA more homogeneous in size, namely, 3×10^6 molecular weight (Qasba *et al.*, 1974). *Empty polyoma capsids* which contain no detectable DNA are also found (Crawford *et al.*, 1962; see also this volume, Chapter 2). Finally, a number of *abnormal particles* are found that are elongated in shape, ranging from oval to long structures (Mattern *et al.*, 1967).

During a lytic polyoma infection, not only is viral DNA synthesized, but the synthesis of host DNA is also stimulated (Dulbecco *et al.*, 1965). As will be shown, some of this newly synthesized host DNA eventually is fragmented to pseudoviral DNA. However, host DNA synthesized before infection is also a source of pseudoviral DNA.

2.1.2. The Evidence for the Induction of Host DNA Synthesis During Polyoma Infection

It is appropriate to consider briefly some aspects of the induction of host DNA synthesis in the context of polyoma pseudovirus biology. The induction of host-cell DNA synthesis is an important phenomenon in the replication of papovaviruses and is extensively covered in the excellent chapter on papovaviruses in these volumes. There are two other excellent reviews dealing with the induction of host DNA synthesis. A comprehensive and well-documented coverage of the subject up to 1969 has appeared (Winocour, 1969). The Cold Spring Harbor monograph edited by Tooze (1973) also has an excellent treatment of the subject.

Minowada and Moore (1963), using autoradiographic techniques, were the first to report that polyoma infection stimulated host-cell DNA synthesis. In a classic paper, Dulbecco *et al.* (1965) showed that polyoma infection of BMK cells caused a tenfold increase in DNA synthesis as compared to mock-infected cells and that most of the DNA synthesized in the infected cells was cellular in nature. These results were rapidly confirmed by a number of groups (Weil *et al.*, 1965; Winocour *et al.*, 1965; Gershon *et al.*, 1965).

There appears to be a general stimulation of all cellular DNA synthesis, not just nuclear DNA, since the synthesis of mitochondrial DNA is also stimulated after polyoma infection of confluent monolayer cultures of mouse cells (Vesco and Basilico, 1971). In the case of the nuclear DNA, much of the newly synthesized DNA contains one parental strand made before infection (Weil *et al.*, 1965; Ben-Porat and

Kaplan, 1967, Kasamaki *et al.*, 1968). Basilico *et al.* (1966), using multiplicities of infection of 0.5–1000 plaque-forming units (PFU) per cell, showed that the extent of infection and the extent of induction of host DNA synthesis depend on the multiplicity of infection.

Whether the induction of host-cell DNA is necessary for polyoma virus replication has not been clearly established. In many ways, the SV40–BSC-1 cell system may be more amenable for the study of the importance of host-cell DNA induction and viral replication in papovaviruses (see Sect. 2.2).

When the DNA of PME-cells is labeled with radioactive thymidine and the cells are infected with polyoma virus after a thorough chase of the radioactive thymidine, analysis of the DNA of the resulting polyoma particles shows that they contain fragments of radioactive cellular DNA but virtually no radioactive polyoma DNA (Qasba and Aposhian, 1971). However, if the radioactive thymidine is added shortly after infection (Yelton and Aposhian, 1972, 1973) at least two types of particles are found. One contains fragments of radioactive cellular DNA and the other contains radioactive polyoma DNA. Thus, the DNA found in pseudovirions appears to contain host DNA synthesized before and after infection. Türler (1974*b*), using 5-bromodeoxyuridine-prelabeled mouse embryo cells, concluded that pseudoviral DNA is excised from both unreplicated and newly replicated regions of mouse DNA after polyoma infection.

2.1.3. Preparation and Purification

Polyoma pseudovirions have been obtained after polyoma infection of BMK cells (Michel *et al.*, 1967; Winocour, 1968), PME cells (Osterman *et al.*, 1970; Qasba and Aposhian, 1971; Aposhian *et al.*, 1972; Qasba *et al.*, 1974), mouse 3T3D cells (Yelton and Aposhian, 1973), and mouse–hamster hybrid cells (Basilico and Burstin, 1971). In my laboratory, virion preparations that consistently contain 80–90% pseudovirions are produced routinely by infecting PME cells. The procedure consists essentially of labeling the host-cell DNA by adding radioactive thymidine to the medium prior to infection with polyoma virus. Immediately before infection, the radioactive medium is changed. The cells are infected and maintained in a medium containing 50 μg of nonradioactive thymidine per ml and 10 μg of nonradioactive deoxycytidine per ml of medium. Since much of the information about polyoma pseudovirions has been obtained using pseudovirions produced in PME cells, this method will be presented in detail.

Mouse embryo cells (2×10^8) in 160 ml of Eagle's medium containing twice the standard amounts of amino acids and vitamins and 10% calf serum (ETC medium) are added to 1-gal flint glass bottles. The bottles are rotated on a roller machine at a speed of 0.3 rpm for 24 hr at 37°C. Then 250 μCi of ^{3}H-thymidine (>15 Ci/mmole) is added to each bottle. Four days after this addition the medium is discarded. The confluent monolayers of each bottle are rinsed twice, using a total of 200 ml of ETC medium containing 50 μg of nonradioactive thymidine and 10 μg of nonradioactive deoxycytidine per ml (ETC, dT, dC medium), and the rinses are discarded. The radioactivity is further chased by incubating for 2 hr with ETC, dT, dC medium. After removing the medium, 15 ml of ETC, dT, dC medium containing polyoma virus are added to each bottle. The multiplicity of infection is 0.1–1 PFU per cell. The bottles are rotated for 2 hr to allow virus adsorption to occur. Then 160 ml ETC, dT, dC medium are added to each bottle and the rotation of the bottles is resumed. Six days post-infection the cells are harvested and then disrupted by cycles of freezing and thawing. The preparation is treated three times with receptor-destroying enzyme (RDE).

Purification is accomplished as follows: the Tris-RDE extracts are centrifuged at 10,000 rpm for 20 min at 20°C, and the pellet is discarded. The supernatant is centrifuged at 27,000 rpm for 3 hr using a Spinco 30 rotor in order to concentrate the virus; the pellets are suspended in CsCl having a density of 1.302 g/ml (the CsCl is dissolved in 0.01 M Tris-HCl, *p*H 7.5, containing 0.06 M KCl), and centrifuged for 22 hr, at 20°C, 32,000 rpm, in a Spinco SW50.1 rotor. After the isopycnic centrifugation, a typical profile, as shown in Fig. 1, is obtained.

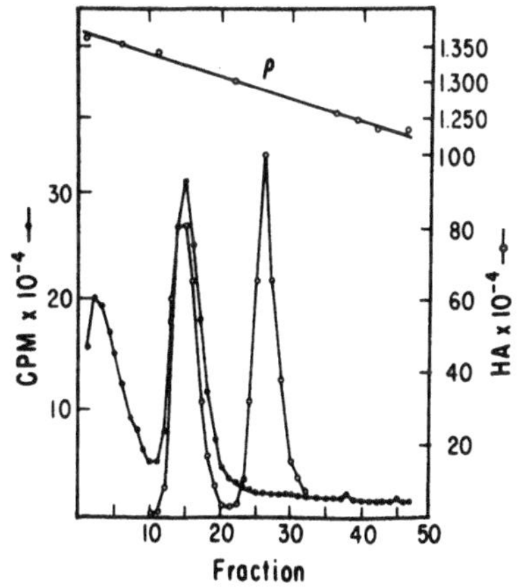

Fig. 1. Equilibrium centrifugation in CsCl of polyoma particles produced in primary mouse embryo cells labeled with ^{3}H-thymidine prior to infection (Qasba *et al.*, 1974).

If pseudovirion preparations containing a minimal number of polyoma virions are desired, the trailing edge (1.310–1.295 g/ml) of the 1.32-g/ml peak is collected and recentrifuged in CsCl. In this manner, pseudovirion preparations can be obtained which contain less than one polyoma particle per 10^6 pseudovirus particles. However, if larger amounts of radioactive pseudovirions are needed, in which the presence of nonradioactive polyoma virus is immaterial, the complete 1.32-g/ml peak may be pooled and rebanded in CsCl. To obtain pseudovirions free of unencapsidated DNA, the preparation after the second CsCl centrifugation should be sedimented in a neutral 10–40% sucrose gradient. This method of extraction and purification, with minor modification, is essentially a combination of parts of the procedure used for polyoma virus as described by Winocour (1963) and Crawford (1969*a*). DNase and RNase are not used in the purification.

Pseudovirions have also been obtained from polyoma-infected BMK cells (Michel *et al.*, 1967; Winocour, 1967*a,b*, 1968). However, it is of interest that Türler (personal communication) has indicated that, in the mouse kidney system, virus isolated from the cell layer always has a higher proportion of pseudovirions than virus isolated from the supernatant medium. This observation should be a note of caution to those who harvest virus only from the supernatant medium in the polyoma–mouse kidney system, as is done routinely in many laboratories.

Density labeling with 5-bromodeoxyuridine has not been helpful in isolating pure pseudovirions in which the bromouracil has substituted for thymine in one strand of the pseudoviral DNA (Türler, 1974*b*; Aposhian, unpublished). This is because the density increment (approximately 0.006 g/ml) is too small to separate them from the virions and pseudovirions which contain unsubstituted DNA.

The mouse–hamster hybrid somatic cells infected with polyoma virus, as reported by Basilico and Burstin (1971), yield complete particles, 90% of which are pseudovirions. This system might be one of the easiest sources of polyoma pseudovirions and deserves more study.

Finally, it should be pointed out that the factor or factors that determine how many pseudovirions are produced in a given virus infection is still relatively unknown. Certainly, the host cell is one of the factors (see Sects. 2.1.6 and 2.2.5). However, Yelton and Aposhian (unpublished) have observed that significantly greater amounts of pseudovirions relative to virions are produced when mouse embryo cells are infected with crude preparations of virus produced by Tris-RDE extraction rather than using the same virus preparations after purification by

CsCl isopycnic centrifugation. Whether this is because the crude preparations contain an inhibitor of polyoma DNA synthesis (viral or nonviral in composition) is unknown at this time.

2.1.4. Characterization

2.1.4a. Evidence that Pseudoviral DNA Consists of Fragments of Mouse DNA

The evidence that the DNA molecules in polyoma pseudovirions are linear, double-stranded fragments of mouse DNA is based on experiments using CsCl equilibrium centrifugation, DNA–DNA hybridizations, DNA–RNA hybridizations, sedimentation analysis in either CsCl or sucrose gradients, melting behavior, and electron microscopy. In most of these studies, advantage was taken of the fact that when the DNA of the host cell is labeled with radioactive thymidine before polyoma infection and steps are taken to remove the radioactive medium and ensure that the specific activity of any remaining radioactive thymidine is sufficiently decreased by a chase before infection with polyoma virus, the particles produced after infection contain fragments of radioactive mouse DNA and do not contain any radioactive polyoma DNA. If radioactive thymidine is added after infection, two types of particles are produced; some contain radioactive polyoma DNA, while others contain fragments of radioactive mouse DNA.

Winocour (1968) extracted the radioactive DNA from particles produced in BMK cells whose DNA had been labeled with ^{14}C-thymidine before infection or with ^{3}H-thymidine after polyoma infection. The DNA from particles produced in cells that were labeled post-infection was further purified using MAK columns to remove components other than type I polyoma DNA. When the labeled DNAs purified from each kind of particle were mixed and centrifuged to equilibrium in CsCl, two peaks of radioactivity were found (Fig. 2). The ^{14}C-DNA had a density equal to mouse DNA (1.702 g/ml) and the ^{3}H-DNA had a density equal to polyoma DNA (1.709 g/ml). Michel *et al.* (1967) and Aposhian (unpublished) also have shown that polyoma pseudoviral DNA has a density equal to mouse DNA.

One of the best methods for showing the DNA in pseudovirions to be mouse DNA and not polyoma DNA is by use of DNA–DNA molecular hybridizations. Qasba and Aposhian (1971) extracted DNA from highly purified pseudovirions obtained from polyoma-infected primary mouse embryo cells that were labeled with ^{3}H-thymidine

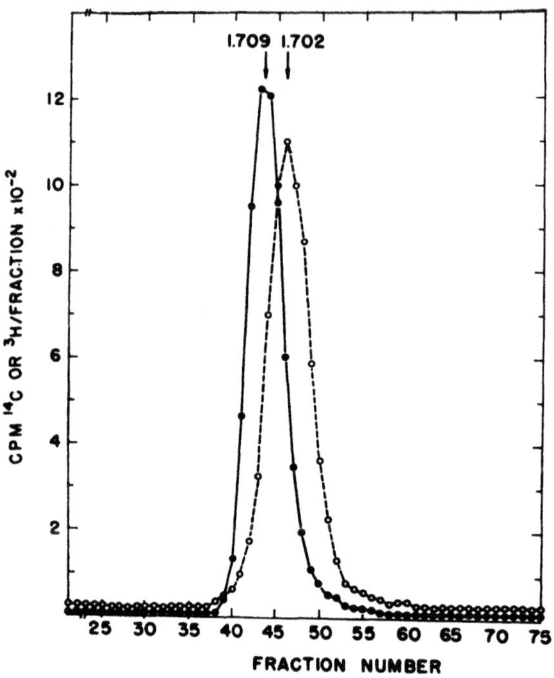

Fig. 2. Buoyant density in CsCl solution of radioactive DNA extracted from particles produced in cells labeled before and after infection. The ^{14}C-DNA (O---O) was extracted from particles produced in cells whose DNA was labeled with ^{14}C-thymidine prior to infection. The ^{3}H-DNA (●——●) was extracted from particles produced in cells labeled with ^{3}H-thymidine after infection and was chromatographed on a MAK column to remove components other than the renaturing 20 S component (Winocour, 1968).

before infection. In order to test the pseudovirus preparation for any contaminating radioactive DNA or polyoma virus, the DNA of the ^{3}H-thymidine-labeled pseudovirions was hybridized with highly purified, polyoma nonradioactive type I DNA. The data in Table 1 show that even with extremely high input counts (177,040 cpm) of ^{3}H-DNA extracted from pseudovirions, no more radioactivity binds to filters loaded with unlabeled polyoma DNA than to blank filters. Under identical conditions, ^{3}H-DNA from polyoma virus hybridizes with unlabeled polyoma DNA with an efficiency of about 50–67%. Thus, the pseudovirus preparations used for analysis are free of radioactive polyoma virus or DNA. The data summarized in Table 2 demonstrate that the DNA extracted from purified pseudovirions hybridizes with mouse embryo DNA. The efficiency of the hybridization of pseudovirus DNA with mouse embryo DNA is comparable to that found when mouse embryo DNA is hybridized with itself.

Similar DNA DNA hybridizations using the DNA from pseudovirions produced in BMK cells have been reported by Winocour (1968).

Base-sequence homology between mouse RNA, synthesized using *Escherichia coli* polymerase, and pseudoviral DNA has also been shown by Winocour (1967*a,b*) and Michel *et al.* (1967). Pulse-labeled RNA extracted from uninfected baby mouse kidney cells also have homology with pseudoviral DNA (Michel *et al.*, 1967).

The above studies, namely, density determination and molecular hybridization, demonstrate that pseudoviral DNA consists of sequences of mouse DNA. That pseudoviral DNA is linear and double-stranded, with a molecular weight of approximately 3×10^6, has been

TABLE 1

Purity of Pseudovirus as Judged by the Failure of Its DNA to Hybridize with Polyoma DNA[a]

Input, cpm	Bound, cpm	Bound to blank filters, cpm	Hybridization	
			cpm	% of input
^{3}H-DNA of pseudovirus				
88,520	475	179	296	0.33
123,928	361	394	<0	<0
177,040	364	551	<0	<0
^{3}H-DNA of polyoma virus (type I)				
1,213	813	0	813	67.0
2,426	1471	3	1468	60.1
4,852	2485	10	2475	51.0
7,278	3521	12	3509	48.2
9,704	5010	47	4963	51.6
^{3}H-DNA of mouse embryo				
84,996	614	150	464	0.54
127,494	536	280	256	0.20
169,992	949	362	587	0.34

[a] Nitrocellulose filters were loaded with 2 μg of sonicated unlabeled, type I DNA of polyoma virus (Qasba and Aposhian, 1971).

TABLE 2

Identification of Pseudovirus by DNA–DNA Hybridization[a]

Input, cpm	Bound, cpm	Bound to blank filters, cpm	Hybridization cpm	Hybridization % of input
Pseudovirus ^{3}H-DNA				
1,872	519	11	508	27.1
3,744	979	18	961	25.6
5,616	1192	36	1156	20.6
7,488	1504	32	1472	19.7
9,360	2038	44	1994	21.3
Mouse ^{3}H-DNA				
2,310	585	28	557	24.1
4,620	1119	86	1033	22.4
5,775	1264	114	1150	22.1
6,930	1533	117	1416	23.9
11,550	2990	225	2765	22.6

[a] Nitrocellulose filters were loaded with 5 μg of mouse embryo DNA (Qasba and Aposhian, 1971).

demonstrated by three different laboratories. When highly purified pseudovirions are prepared from polyoma-infected BMK cells, 3T3D mouse cells, or PME cells, and when the DNAs of these pseudovirions are extracted and examined by sedimentation through neutral sucrose gradients, the DNA from the pseudovirions cosediments with a 14 S marker (Yelton and Aposhian, 1973). Thus, the pseudoviral DNAs derived from particles produced in three different cell types have an apparent sedimentation value of 14 S in neutral sucrose. Two other laboratories have shown that the DNA derived from particles produced in BMK cells sediments at approximately 14.6 S (Winocour, 1968) or <14 S (Michel *et al.*, 1967). The reasons for obtaining S values of less than 14 S by one of the laboratories are unexplained.

Michel *et al.* (1967) analyzed, by a number of methods, a polyoma virus preparation that had been purified by CsCl equilibrium centrifugation plus sedimentation through a sucrose gradient. The DNA was then extracted and examined by velocity-band centrifugation. A fast (20 S) and a slow (~14 S) band were obtained. The 20 S fractions consisted of twisted circular DNA with a mean length of 1.75 μm, as expected for polyoma type I DNA. Analysis by electron microscopy of the 14 S fractions showed 95% of the DNA molecules to be linear with a very broad length distribution. Purified pseudovirions of the lightest density contained only linear DNA molecules. As the buoyant density of the pseudovirions decreased, the mean length of the DNA extracted from them was found to decrease. Pseudovirions that were found in the region of the empty capsids contained linear helices that had a length of one-fifth or less the length of circular polyoma DNA. The length distributions of linear molecules were very broad and no molecules longer than circular polyoma DNA were found (Fig. 3).

The investigations summarized in this section indicate that polyoma pseudoviral DNA consists of fragments of mouse DNA, is linear, double-stranded, and has an approximate molecular weight of 3×10^6.

2.1.4b. Evidence that Pseudoviral Coats are Polyoma Coats

Winocour (1968) has compared the abilities of polyoma virus and pseudovirus particles to bind to red blood cells of three different species. Polyoma virus, when mixed with red blood cells, causes a response depending on the species of red cells. The virus strongly agglutinates guinea pig red blood cells, partially agglutinates rat red blood cells, and does not agglutinate horse red blood cells. In the experiment summarized in Table 3, radioactive particles produced in cells labeled with ^{3}H-thymidine after or before infection with polyoma virus were mixed with the red cells from the various species. After sedimentation of the cells, the radioactivity remaining in the supernatant was determined. A control experiment using sonicated, uninfected mouse kidney cells labeled with ^{3}H-thymidine was also performed. The results of these hemagglutinating experiments (Table 3) demonstrate that the pseudovirus particles (particles labeled before infection) respond in the same way as those of normally labeled polyoma virus. Winocour (1968) found also that pseudovirions do not bind to guinea pig red cells at 37°C and that when pseudovirions are mixed with guinea pig cells at 4°C and then incubated at 37°C, the virus is eluted from the red cells. Such behavior is also characteristic of normal polyoma virus.

2.1.5. Intracellular Events Related to Production

Some of the intracellular events related to polyoma pseudovirion production in PME cells have been investigated by Yelton and Aposhian (1972). The cells were infected with polyoma virus, and ^{3}H-thymidine was added immediately after the 2-hr adsorption period. Poly-

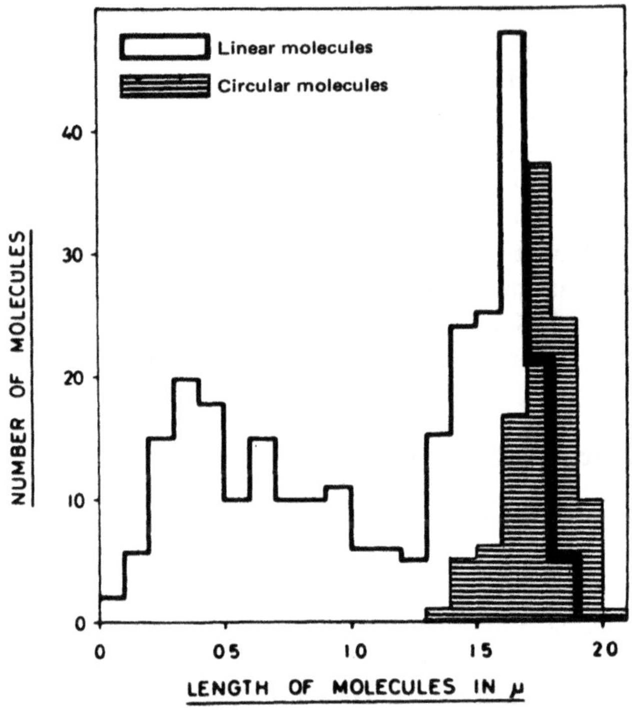

Fig. 3. Histogram of the lengths of circular and linear polyoma DNA molecules. DNA from a viral preparation containing 25% type I, 5% type II, and 70% type III DNA (Michel *et al.*, 1967).

oma DNA synthesis and production of host-cell 16 S DNA were measured by sedimenting the 0.6% SDS–1 M NaCl supernatant, prepared by the method of Hirt (1967), through alkaline sucrose gradients. Virus particles were obtained from the infected cells by using the RDE methods of Crawford (1969*a*). The particles were purified by CsCl equilibrium centrifugation, and the amounts of complete virions and empty particles in the peaks having a density of 1.32 and 1.28 g/ml were determined by measuring the hemagglutination of guinea pig red cells; PFU was also measured.

Replication of polyoma DNA was found to begin by 18 hr after infection and was completed by 48 hr (Fig. 4). Assembly of polyoma capsids began by 30 hr post-infection, as measured by the increase of empty capsids. Intracellular fragments of host-cell DNA of the size found in pseudovirions were first detected 36 hr after infection. Production of full particles began by 42 hr post-infection and continued until 96 hr.

The time course of polyoma replication in PME cells was similar to that found by others using a variety of cell lines (Ben-Porat and Kaplan, 1967; Dulbecco *et al.*, 1965; Gilead, 1972; Kit *et al.*, 1966). However, a large amount of 14 S host DNA fragments was found to be produced in the PME cells.

It should be emphasized that significantly more radioactivity was associated with the 14 S host DNA fragments than was associated with the polyoma DNA molecules. Because the specific activities of the 14 S DNA and the polyoma DNA were about the same, the ratio of their radioactivities gives an indication of the relative quantity of each produced. When full particles were first detected (42 hr), the ratio of 14 S DNA to polyoma DNA was about 3:1; by 96 hr it was 7:1. The production of 14 S DNA in amounts much greater than polyoma DNA appears to be unique to PME cells; it was not found after infection of BMK or 3T3D cells (Yelton and Aposhian, 1973).

Not only was a large quantity of 14 S DNA produced intracellularly, but this DNA appeared to be very homogeneous in size (Yelton and Aposhian, 1972). The reason for this homogeneity is unknown. It may be that the host-cell DNA is fragmented into a variety of sizes which nucleases further degrade unless the fragment is protected. Be-

TABLE 3

Binding to Red Blood Cells of Radioactive Particles Produced in Cells Labeled with ^{3}H-Thymidine Before and After Infection[a]

	Percentage of initial radioactivity remaining in supernatant after sedimentation of red blood cells at 4°C		
Origin of red blood cells	Radioactive particles from cells labeled after infection	Radioactive particles from cells labeled before infection	Sonicate of uninfected, ^{3}H-thymidine-labeled mouse cells
Guinea pig	8.3	5.7	69.3
Rat	49.4	47.3	71.6
Horse	68.8	67.0	70.2

[a] The radioactive particles from cells labeled after infection are predominantly polyoma virions, while those from cells labeled before infection are pseudovirions. (Winocour, 1968).

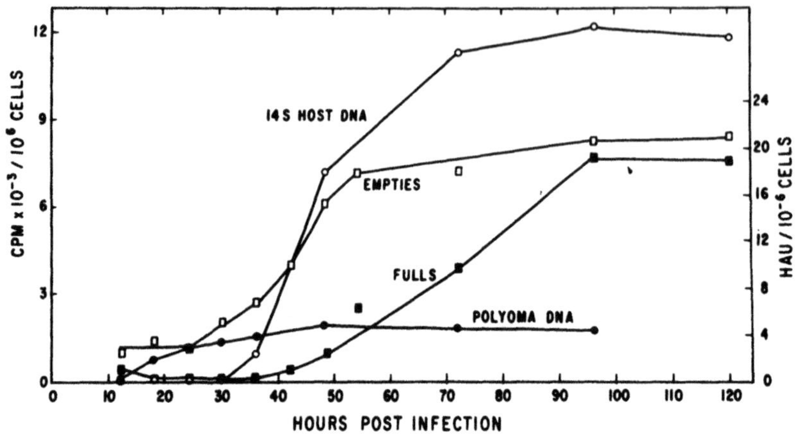

Fig. 4. Relationship of the intracellular events leading to the production of pseudovirions. Infected cells, at the indicated times, were analyzed for polyoma DNA, 14 S host-cell DNA fragments, full particles, and empty particles. Polyoma DNA synthesis and production of 14 S host-cell DNA fragments were measured by counting the radioactivity in the 53 S and 16 S peaks after sedimentation through alkaline sucrose. As a label, ^{3}H-thymidine was added immediately after infection (Yelton and Aposhian, 1972).

cause 14 S fragments are about the same size as polyoma DNA, this protection from nucleases might occur by the fragments interacting with a viral protein.

Although others have reported that the percentage of pseudovirions in polyoma virus preparations from BMK cells varies from 10 to 90%, the usual value is about 10%. With the PME cells, however, it has been found consistently in this laboratory that 80% or more of the full particles are pseudovirions. This large percentage of pseudovirions may be related to the large amount of 14 S host DNA fragments produced during infection of PME cells. Because the production of infectious virus mirrors the production of pseudovirions (Yelton and Aposhian, 1972), there appears to be no mechanism for selectively encapsidating polyoma DNA. Rather, any intracellular DNA molecules of appropriate size seem to be packaged. Hence, the relative pool sizes of polyoma DNA and 14 S host DNA at the time of virus assembly may dictate the amounts of polyoma virus and pseudovirus assembled. However, this may be an oversimplification of a complex series of interacting events.

Türler (personal communication) has also studied the kinetics of pseudovirion formation in relation to the kinetics of the synthesis of mouse DNA, polyoma DNA, mouse DNA fragments, and virions in BMK cells infected with polyoma virus. The results were similar to those in mouse embryo cells reported by Yelton and Aposhian (1972).

The partial inhibition of cellular and viral DNA synthesis by either 5-fluorodeoxyuridine or mitomycin C leads to as much as a threefold increase in the proportion of pseudovirions in the lysates of infected cells. The absolute amounts of pseudovirions and virions, however, were found to be decreased (Türler, personal communication).

Türler (1974*a*) has shown that when mouse embryo cells are prelabeled with 5-bromodeoxyuridine, in the presence of 5-fluorodeoxyuridine, and then infected with polyoma virus in the absence of the bromo analogue, a lytic infection results that is essentially the same as that in cells not exposed to the 5-bromodeoxyuridine. In both cases, the time course of the infection and production of progeny virus were essentially the same. Furthermore, extraction of the DNA from a purified virus preparation containing both polyoma virions and pseudovirions (density 1.315–1.34 g/ml) that had been obtained by infection of density-labeled mouse embryo cells gave a peak of unsubstituted (LL) DNA and a peak of hybrid (HL) DNA. In the case of the HL DNA, bromouracil was substituted for thymine in one strand. The HL peak was shown to consist of virtually pure HL pseudoviral DNA (therefore, fragments of mouse DNA) by the criteria of electron microscopy and the labeling conditions used (Türler, 1974*b*). Comparison of HL and LL polyoma DNA type III from six different virus preparations indicated an excess of HL DNA, leading Türler (1974*b*) to conclude that pseudoviral DNA is excised both from unreplicated and newly replicated regions of the mouse chromosomes.

2.1.6. Influence of the Host Cell

Experiments by Yelton and Aposhian (1973) have indicated that the type of host cell infected by polyoma virus appears to influence the relative amounts of polyoma virions and pseudovirions that are produced. Three mouse cell types permissive for polyoma infection were used; PME, BMK, and mouse 3T3D cells were infected at a multiplicity of infection of 0.5 PFU per cell. Medium containing ^{3}H-thymidine was added after the 2-hr virus adsorption period. The virus yields were harvested at 96 hr post-infection and purified by centrifugation in CsCl and sedimentation through sucrose. The purpose of the sedimentation through sucrose is to remove any DNA adsorbed to or trapped on the external surface of the virions. The purified virus particles were lysed in 0.3 N NaOH and their DNA analyzed by sedimentation through alkaline sucrose gradients (Fig. 5). By comparing the amounts of radioactivity in the 53 S and 16 S peaks, the relative amounts of polyoma virions and pseudovirions can be determined. In the case of pseudovirions, the radioactivity represents pseudovirus produced from cellular DNA labeled after infection. The reader is reminded that pseudoviral DNA sediments at 14 S in neutral gradients, but at 16 S in alkaline gradients.

Pseudovirus represented 20, 38, and 69% of the full particles produced in 3T3D, BMK, and PME cells, respectively (Fig. 5). That the measurements of radioactivity in the 53 S and 16 S areas of such alkaline sucrose gradients are a valid method of determining the polyoma and pseudovirus content of purified radioactive polyoma preparations was demonstrated by DNA–DNA hybridization (Yelton and Aposhian, 1973). By the hybridization method, pseudovirus represented 20, 37, and 63% of the full particles produced in 3T3D, BMK, and PME cells, respectively.

The amount of polyoma pseudovirus produced is influenced also by the length of time that infection is allowed to proceed in particular cell types. When the particles harvested from the BMK cells at different times after infection were analyzed by sedimentation in alkaline sucrose gradients, the relative proportion of pseudovirus produced increased from 16 to 38% between 48 and 96 hr post-infection (Table 4). For 3T3D cells, the proportion increased from 15 to 20%, and for PME cells from 56 to 69% (Table 4).

Yelton and Aposhian (1973) also determined the amount of host-cell 16 S DNA fragments produced during infection of each of the cell types. The low-molecular-weight DNA was extracted from the cells at 36, 48, and 60 hr post-infection using the method described by Hirt (1967) and analyzed by sedimentation in alkaline sucrose gradients. Following polyoma infection, some fragmentation of host-cell DNA was found in the BMK and 3T3D cells, but it was not as extensive as that seen with the PME cells (Table 5). In the PME cells, by 60 hr post-infection, 67% of the intracellular low-molecular-weight DNA sedimented at 16 S and was of cellular origin.

Since extensive fragmentation of host-cell DNA is seen after infection of the PME cells, but not the BMK or 3T3D cells, Yelton and Aposhian (1973) investigated pseudovirus production in mouse embryo cells prepared from mice obtained from three different supply houses. Extensive fragmentation of host-cell DNA following polyoma infection was found regardless of the source of the mice. Therefore, the strain of

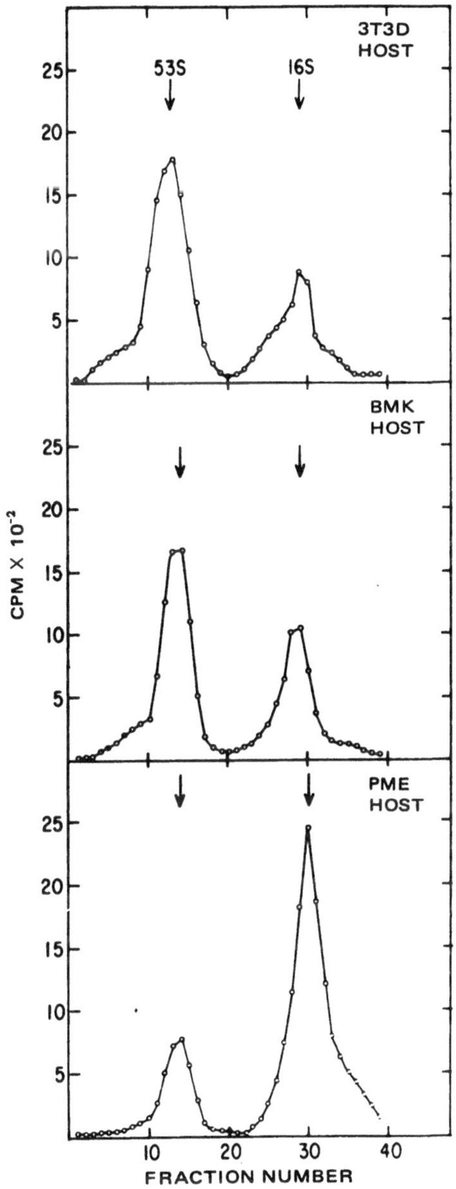

Fig. 5. The 53 S and 16 S DNA in virus particles purified from the different cell types 96 hr after infection with polyoma virus. ^{3}H-thymidine was added after the 2-hr adsorption period. The purified virus particles were lysed in 0.3 N NaOH and layered onto alkaline 5–20% sucrose gradients. The arrows represent the positions of ^{14}C-DNA markers (Yelton and Aposhian, 1973).

TABLE 4

Relative Amounts of Polyoma DNA and Pseudovirus DNA Found in Virus Particles Harvested at Different Times Post-Infection[a]

Cell type	Time of harvest, hr	Polyoma DNA, %	Pseudoviral DNA, %
BMK	48	84	16
	72	70	30
	96	62	38
3T3D	48	85	15
	72	84	16
	96	80	20
PME	48	44	56
	72	32	68
	96	31	69

[a] The cells were infected at a multiplicity of infection of 0.5 PFU/cell and labeled with ^{3}H-thymidine immediately following the 2-hr adsorption period (Yelton and Aposhian, 1973).

mice from which PME cells are prepared does not appear to affect the relative amounts of polyoma DNA and 16 S host DNA fragments formed in the cells.

In the experiments of Yelton and Aposhian (1973), infected BMK cells showed the greatest variability in the proportion of pseudovirus

TABLE 5

Relative Amounts of Polyoma DNA and Fragmented Host-Cell DNA (16 S) Found Intracellularly at Different Times Post-Infection[a]

Cell type	Time of harvest, hr	Polyoma DNA, %	Host-cell DNA, %
BMK	36	97	3
	48	94	6
	60	82	18
3T3D	36	98	2
	48	98	2
	60	94	6
PME	36	90	10
	48	52	48
	60	33	67

[a] Yelton and Aposhian (1973).

produced. In five different experiments, the proportion varied from 26 to 54%. Michel *et al.* (1967) also found wide variation in the relative amounts of pseudovirus produced in BMK cells. This variability may be related to the number of fibroblastic cells in the culture. Primary cultures of BMK cells consist mainly of epitheliallike polygonal cells and a small amount of clongated fibroblastic cells. After a sheet of confluent cells is formed, the DNA-synthesizing activity of the culture is very low. The small number of DNA-synthesizing cells in such cultures is confined to the fibroblastic type of cells. Pseudovirus production may be related to the fibroblast content of a given cell population.

In the experiments of Yelton and Aposhian (1973), when BMK cells were used, initially almost all the cells were epithelial-like. However, as the infection proceeded, the epithelial-like cells became round in their shape and became detached from the surface of the Petri dishes. Concurrently, fibroblastic cells became more apparent and increased in number. The earlier the appearance of fibroblastic cells, the greater was the final proportion of pseudovirions.

Mouse embryo cells are predominantly fibroblastic and produce the greatest proportion of pseudovirions after infection. If fibroblastic cells are indeed responsible for or related to pseudovirus production, why don't 3T3D cells produce more pseudovirions after polyoma infection? The 3T3D fibroblasts may have lost the required property for pseudovirus production during the original selection process.

Basilico and Burstin (1971) have reported a mouse–hamster hybrid cell line which produces almost entirely pseudovirions following polyoma infection. Even though no intracellular pool of fragmented host cell DNA was found in these hybrid cells, 90% of the complete particles were polyoma pseudovirions. Unlike the PME cells, however, in which polyoma DNA replication is normal, the mouse–hamster hybrid cells greatly restricted the replication of polyoma DNA.

In PME, BMK, and 3T3D cells, induction of host-cell DNA synthesis as well as its excision or fragmentation must occur following infection with polyoma virus. When ^{3}H-thymidine is added after virus adsorption, not only is the polyoma DNA labeled but also the host-cell DNA becomes radioactive, indicating that an induction of host-cell DNA synthesis occurred in the confluent monolayers. When the host-cell DNA is radioactively labeled before polyoma infection, fragments of this radioactive cellular DNA are found in the pseudovirions produced during infection. Thus, a process of excision or fragmentation is active in each of the cell types. However in the case of the BMK cells it is also possible that the epitheliallike cells lack an excision mechanism and therefore do not produce pseudovirions. The fibroblastic cells in the population of BMK cells, however, have the excision process and therefore produce pseudovirions. This hypothesis has not as yet been put to a rigorous test.

2.1.7. Biological Properties

In 1968, a research program was started in my laboratory to investigate the biological properties of polyoma pseudovirions. The aim of the program was to attempt the development of a DNA and gene delivery system for cultured mammalian cells by using polyoma pseudovirions. Six questions were proposed as a framework for a step-by-step investigation of the problem.

1. Can radioactive pseudovirus free of polyoma virus be prepared?
2. Does radioactive pseudovirus penetrate the cell?
3. Is pseudovirus uncoated?
4. Is pseudoviral DNA found in the nuclei of cells infected with polyoma pseudovirions?
5. Is pseudoviral DNA incorporated into the mouse genome?
6. Is there a phenotypic expression of pseudovirus?

We were indeed fortunate to have chosen the PME system for polyoma pseudovirus production since this system consistently produces during infection the most pseudovirions relative to polyoma virions. Highly purified radioactive polyoma pseudovirions, containing a minimum of infectious polyoma virions and virtually no radioactive polyoma virions can be obtained with relative ease (Osterman *et al.*, 1970; Qasba and Aposhian, 1971).

2.1.7a. Adsorption, Penetration, and Uncoating

Polyoma pseudovirions are absorbed to and uncoated by secondary mouse embryo (SME) cells (Osterman *et al.*, 1970). By 90 min post-infection, approximately 75% of the virus has been adsorbed (Table 6). No attempt is made to cool the cells during the adsorption process or during the time the inoculum is removed and the cells harvested. This is probably the cause for the small amount of adsorbed radioactivity at 0 time.

The first experiments to determine if polyoma pseudovirions are

TABLE 6

The Uncoating of Pseudovirus in Secondary Mouse Embryo Cells as Measured by Susceptibility to Pancreatic DNase[a]

	Amount of time cells were exposed to virus		
	0 min	60 min	90 min
	cpm	cpm	cpm
Expt. 1			
Pseudovirus adsorbed	1,470	22,300	26,500
Pseudovirus uncoated[b]			
− DNase	5	30	75
+ DNase	85	915	1,560
Expt. 2			
Pseudovirus adsorbed	3,750		25,200 25,600
Pseudovirus uncoated			
− DNase	0		55 70
+ DNase	140		985 930

[a] Purified pseudovirus (36,200 cpm/plate and 33,000 cpm/plate for Expts. 1 and 2, respectively) were added to each plate at 0 time.

[b] Measured as a conversion by pancreatic DNase of acid-insoluble radioactivity (Osterman *et al.*, 1970).

uncoated in mouse cells made use of pancreatic DNase. Radioactive pseudoviral DNA that is uncoated is susceptible to the action of pancreatic DNase; the radioactivity will be converted from an acid-insoluble to acid-soluble form. Encapsidated DNA, however, is protected from the DNase and is not converted from an acid-insoluble to an acid-soluble form by action of the enzyme. Any encapsidated DNA remains acid insoluble. By 90 min, approximately 6% of the adsorbed pseudovirions are uncoated (Table 6). The small increase in acid-soluble radioactivity (5 to 75 cpm) in the absence of *added* pancreatic DNase is probably due to the action of a small amount of endogenous DNase in the cell extract.

The appearance of radioactivity in the nuclear fraction of SME cells after infection with ^{3}H-thymidine-labeled polyoma pseudovirions offers further evidence that polyoma pseudovirions can enter mouse cells (Qasba and Aposhian, 1971). The kinetics of the appearance of polyoma virions or pseudovirions in the cytoplasmic or nuclear fraction of SME cells is shown in Fig. 6. The data are expressed as percentages of the total radioactivity found in the cell, i.e., as a percentage of the sum of the radioactivity in the cytoplasmic and nuclear fractions. There is an extremely rapid increase of the radioactivity (reaching a peak in about 30 min) in the cytoplasmic fraction after either pseudovirus or polyoma infection (Fig. 6). The amount of radioactivity found in the nuclear fraction is small, but it steadily increases. In the case of pseudovirus, 24% of the total radioactivity found in the cell 24 hr post-infection is in the nuclear fraction. This represents about 7.1% of the

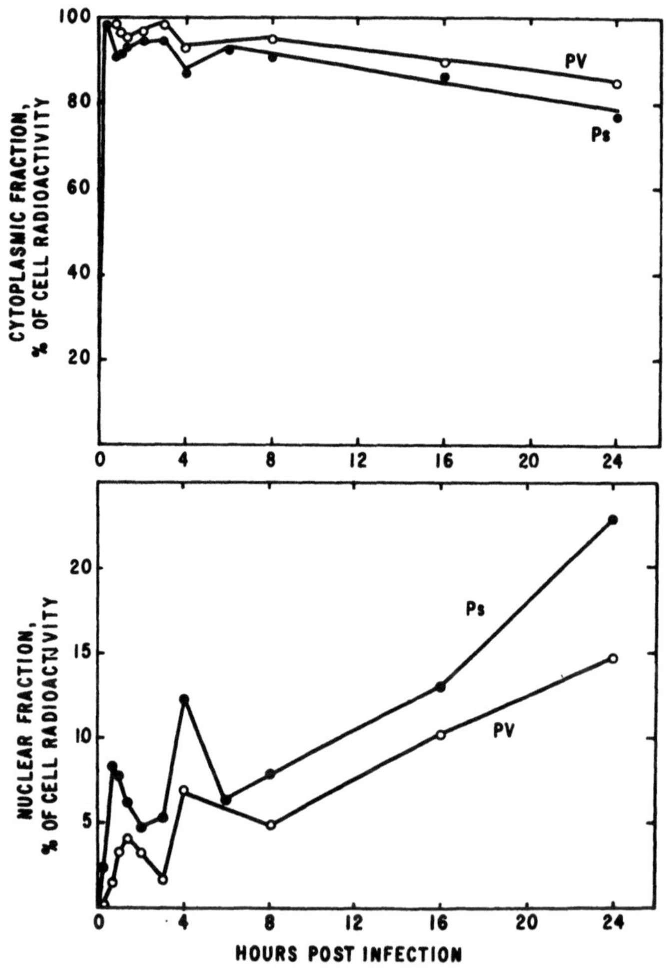

Fig. 6. Appearance of pseudovirus (Ps) or polyoma virus (PV) radioactivity in cytoplasmic and nuclear fractions of SME cells. ^{3}H-thymidine-labeled virus was added to each plate of confluent SME cells (4×10^6 cells per 60-mm Petri dish). For pseudovirus infection, 4×10^3 hemagglutinating units and 1.4×10^4 cpm per plate were used. For polyoma infection, 1.6×10^3 hemagglutinating units and 9.85×10^3 cpm per plate were used (Qasba and Aposhian, 1971).

TABLE 7

DNase Sensitivity of the Radioactivity Found in the Nuclei of Secondary Mouse Embryo Cells after Infection with ^{3}H-Thymidine-Labeled Pseudovirus[a]

	% Radioactivity after treatment with DNase	
Fraction	TCA soluble	TCA insoluble
Whole nuclei	68	32
Homogenized nuclei		
Supernatant	53	47
Pellet	78	22

[a] Qasba and Aposhian (1971).

original pseudoviral counts to which the cells were exposed. The initial rapid increase of radioactivity in the nuclear fraction (1–1.5 hr after infection) is believed to be artifactual, as documented by Qasba and Aposhian (1971).

Further proof of the presence of uncoated pseudoviral DNA in cells infected with pseudovirus is indicated in Table 7. Treatment of nuclear preparations of pseudovirus-infected mouse embryo cells with pancreatic DNase demonstrated that 53–78% of the pseudoviral DNA radioactivity is sensitive to the enzyme. The nuclei were homogenized gently so that only the nuclei and not the virions were disrupted.

Radioactivity is found also in primary human embryonic cells infected with ^{3}H-thymidine-labeled pseudovirions. When cytoplasmic and nuclear fractions of infected cells were prepared, 7.3% of the total counts found in the human cells were associated with the nucleus (Table 8). The radioactivity in the nuclear fraction can be shown to represent uncoated pseudoviral DNA by gently homogenizing the nuclei from such infected cells and analyzing the supernatant by sedimentation through neutral sucrose (Fig. 7). Two peaks of radioactivity are found. One cosediments with intact ^{14}C-labeled pseudovirus used as a marker; the other represents uncoated pseudoviral DNA. The evidence indicates that polyoma pseudovirus enters human embryonic cells, is associated with the nuclear fraction, and is uncoated. Similar studies have shown that polyoma pseudovirions are uncoated in primary chicken embryo cells (Kashmiri and Aposhian, unpublished).

TABLE 8

Distribution of ^{3}H-Pseudovirus Radioactivity in Human Embryonic Cells after Infection with ^{3}H-Thymidine-Labeled Pseudovirus[a]

		Total cellular radioactivity	
Time, hr	Input cpm bound to or in cells, %	Cytoplasmic fraction, %	Nuclear fraction, %
0	0.8	99.3	0.7
24	21.9	92.7	7.3

[a] Qasba and Aposhian (1971).

2.1.7b. Fate of Pseudoviral DNA

The adsorption of polyoma pseudovirions to mouse and human embryo cells, their uncoating in such cells, and the detection of the uncoated pseudoviral DNA in the nuclear fraction of these cells after infection have been established (Aposhian *et al.*, 1973; Osterman *et al.*, 1970; Qasba and Aposhian, 1971). It is, therefore, not surprising that polyoma pseudovirions have been suggested as potential agents for mammalian gene transfer or transduction (Aposhian, 1970; Hirt and Widmer-Favre, 1972; Michel *et al.*, 1967).

Stable transduction of mammalian cells by pseudovirions would require not only the delivery and uncoating of pseudoviral DNA in the cell but also the integration of part or all of the donor pseudoviral DNA into the DNA of the recipient cell. Generalized transductions of *Salmonella typhimurium* by the generalized transducing particles of P22, for example, result in the physical association of 12–15% of the donor transducing DNA with the DNA of the recipient bacteria, as shown by the elegant experiments of Ebel-Tsipis *et al.* (1972*b*). However, only about 2% of the donor DNA is integrated as large fragments of donor DNA.

What is the fate of polyoma pseudoviral DNA? Is it integrated into the host-cell chromosomes? Is it degraded to precursors of DNA that are reused for DNA synthesis? Or do both phenomena occur?

These questions have been investigated by Kashmiri and Aposhian (1974). When mouse cells are infected for 24 hr with ^{3}H-thymidine-labeled polyoma pseudovirus and the infected cells are then lysed gently on the top of a sucrose gradient, centrifugation of the lysate through the sucrose gradient shows that a substantial percentage of the input pseudoviral radioactivity is recovered in the cellular DNA region of the gradient (Table 9 and left panels of Fig. 8). Mouse cells, depending on their type, vary as to the percentage of input pseudoviral radioactivity found in the cellular DNA region of the gradients. In the case of SME cells, as much as 6.2% of the input counts are found in the cellular DNA region. The 3T3D cells appear to be the most efficient in this respect; as much as 11% of the input pseudoviral counts are recovered in the cellular DNA portion of the gradient. When BMK cells are used, only 0.6% of the input counts is found in the host DNA area.

The amount of radioactivity recovered in the DNA of the recipient cells infected with ^{3}H-thymidine pseudovirus is greatly reduced when medium containing nonradioactive thymidine and deoxycytidine is ad-

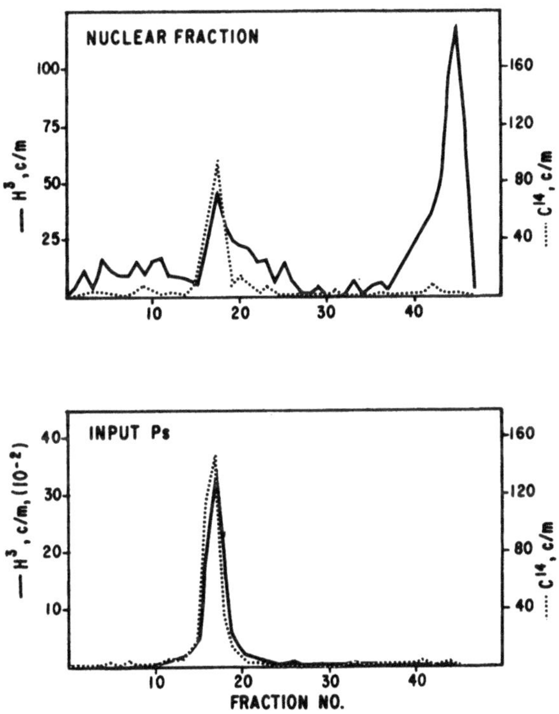

Fig. 7. (Top) Sedimentation in neutral sucrose of disrupted nuclei of human embryonic cells infected with ^{3}H-thymidine-labeled polyoma pseudovirus. As a sedimentation marker, ^{14}C-labeled pseudovirions were added prior to centrifugation. (Bottom) Sedimentation in neutral sucrose of the ^{3}H-thymidine-labeled polyoma pseudovirions used for infection and marker ^{14}C-labeled pseudovirions (Qasba and Aposhian, 1971).

TABLE 9

Radioactivity in Mouse Cell DNA and Pseudovirus DNA after Infection of Mouse Cells with ^{3}H-Thymidine-Labeled Pseudovirus[a]

		Input		Radioactivity in cells as % of input			
				Without chase		With chase[c]	
Cell type (1)	Expt. No. (2)	Pseudo-virions per cell[b] (3)	Counts per min per plate (4)	Cellular DNA (5)	Pseudo-viral DNA (6)	Cellular DNA (7)[d]	Pseudo-viral DNA (8)
SME	1	5,500	32,400	3.0	7.3	0.49 (159)	7.8
	2	27,000	34,800	6.2	16.4	0.30 (105)	12.6
	3[e]	5,500	65,900	1.2	13.0	0.15 (99)	12.1
3T3D	4	14,200	87,400	11.2	10.0	NP	NP
	5	15,500	35,400	6.9	8.8	0.29 (103)	9.3
	6	20,500	42,700	10.5	8.3	0.62 (265)	7.1
BMK	7	20,500	66,500	0.6	3.8	0.11 (73)	4.4

[a] Kashmiri and Aposhian (1974).
[b] Pseudovirus preparations had PFU/HU ratios of 8.0×10^2–1.4×10^3. Polyoma virus stocks prepared in this laboratory from plaque purified virus have a PFU/HU ratio of 5×10^5.
[c] Concentration of thymidine and deoxycytidine were 50 μg and 10 μg/ml, respectively, except for Expt. 6, where they were 250 μg and 50 μg, respectively. NP = not performed.
[d] The numbers in parentheses in this column are the cpm corrected for background.
[e] Spent medium was returned to the cells after pseudovirus adsorption period.

ded after pseudovirus adsorption (Fig. 8 and Table 9). A small number of counts (ranging from 99 to 265 cpm) cannot be chased completely when SME cells or 3T3D cells are the recipients. These counts persist in one of the experiments with 3T3D cells (see Table 9, Expt. 6) even though the concentration of the thymidine and deoxycytidine chase is increased fivefold.

DNA synthesis in mouse cells can be decreased by placing used medium, rather than fresh medium, on the cells after virus adsorption (Todaro *et al.*, 1965). Only 1.2% of the input pseudoviral radioactivity is incorporated into the cellular DNA of SME cells when the used medium is added back to the cells after pseudovirus adsorption (see Table 9, Expt. 3). When used medium plus a chase is used, only 0.15% of the input counts is incorporated into the cellular DNA (see Table 9, Expt. 3). Reconstruction experiments have indicated that the incorporation of

0.15% of the input counts is of doubtful significance (Kashmiri and Aposhian, unpublished).

Furthermore, the addition of cytosine arabinoside, an inhibitor of DNA synthesis, immediately after pseudovirus adsorption, also decreases to almost zero the amount of input pseudoviral radioactivity that is recovered in the cellular DNA (Kashmiri and Aposhian, 1974).

The experiments of Kashmiri and Aposhian (1974) indicate that in mouse cells infected with polyoma pseudovirions much of the pseudoviral DNA is degraded and reutilized for host DNA synthesis. Whether this is true if pseudovirions are used to infect synchronized

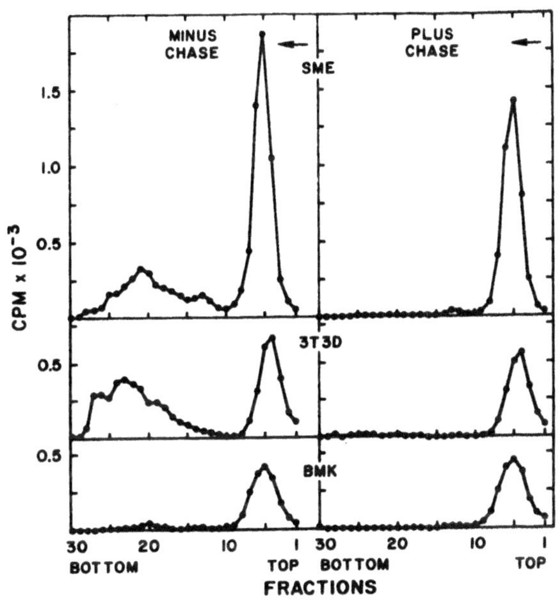

Fig. 8. Appearance of radioactivity in the cellular DNA region of alkaline sucrose gradients after infection of mouse cells with ^{3}H-thymidine-labeled polyoma pseudovirions in the absence (left panels) or presence (right panels) of a thymidine and deoxycytidine chase (Kashmiri and Aposhian, 1974). When a chase was used, 50 μg of thymidine and 10 μg of deoxycytidine were added per ml of medium immediately after the virus adsorption period, except for the 3T3D experiment where the deoxynucleoside concentrations were fivefold greater. The direction of sedimentation is indicated by the arrow. Each fraction of the gradient was precipitated with TCA, filtered, and counted as described by Kashmiri and Aposhian (1974).

cells in different stages of the cell cycle remains to be seen. The meaning of the small number of persistent, nonchaseable radioactive counts that appear in the host cell DNA in some of the experiments is not clear. Kashmiri and Aposhian (1974) have offered various possible interpretations of the meaning of these nonchaseable counts. It thus appears that the question as to whether or not polyoma pseudoviral DNA is integrated into recipient mouse-cell DNA must await other types of experiments using more sensitive physicochemical or genetic techniques.

2.1.7c. Transduction Attempts

As yet, there has not been a successful transfer of an animal cell gene from one animal cell to another via an animal virus. No mammalian cell–animal virus system has been found that is completely analogous to the generalized transduction systems of bacteria and phage. True, in the animal cell systems, the synthesis of new proteins has been directed by viral genes introduced into the cell. Experiments with rabbit papilloma virus (Rogers and Moore, 1963) and ultraviolet-inactivated herpes simplex virus (Munyon *et al.*, 1971) are examples of the viral genome bringing into the cell information for a virus-coded enzyme. However, these are not analogous to transductions.

Hirt and Widmer-Favre (1972) have reported that attempts to transduce the thymidine kinase (TK) gene and the inosinic pyrophosphorylase (IPP) gene using polyoma pseudovirions harvested from primary mouse kidney cells have been unsuccessful. Gene transfer was tested by autoradiography after labeling with tritiated thymidine or hypoxanthine as well as by colony formation in medium restrictive for deficient cells. They estimated that one TK or IPP gene could be introduced into approximately 1 cell out of 100. They observed no gene expression in the 10^7 cells scanned.

Axelrod and Trilling (1972) have prepared polyoma pseudovirions by infecting SV40-transformed mouse cells with polyoma virus. Such pseudovirions were used to infect primary African green monkey cells at a particle-cell multiplicity of 2×10^6 particles per cell. The pseudovirus-infected cells were tested by immunofluorescence for the SV40 T antigen. These attempts to transduce the gene for SV40 T antigen were negative.

When calculations are made about the possible number of cells that might be transduced by pseudovirions, the estimations are not very encouraging. For example, polyoma pseudoviral DNA has a molecular weight of approximately 3×10^6. If 10^{12} is assumed as the approximate molecular weight of the mouse genome, 3×10^5 pseudovirions have an amount of DNA approximately equivalent to the amount of DNA in the mouse genome. If mouse cells are infected with pseudovirions at a multiplicity of 3×10^5 particles per cell, every cell might be expected to receive the desired gene. However, as suggested by the data of Kashmiri and Aposhian (1974), less than 0.5% of the input pseudoviral DNA appears to be associated with host-cell DNA. The absolute figure may be much less than 0.5%. Also, at a multiplicity of infection of 3×10^5 pseudovirus particles per cell, the amount of infectious polyoma virus contaminating some pseudovirus preparations is high enough to cause a lytic infection if mouse cells are used as the host, thereby decreasing the number of surviving cells. However, it is possible to overcome the latter complication by using hamster cells, which are not permissive for polyoma virus.

These negative results using pseudovirions to transduce animal-cell genes are presented to make the reader aware that several attempts have been made. Most investigators do not know whether the negative results in these transduction attempts are due to unknown technical problems or not.

In keeping with this effort to make others aware of transduction attempts in this important area of investigation, the following experiments may be of interest to the reader. Ottolenghi-Nightingale (1969) has transformed white mouse skin cells, which do not produce melanin, to melanin-producing cells by incubating white mouse cells with DNA prepared from black mice. An attempt was made to extend the experiments of Ottolenghi-Nightingale by infecting PME cells prepared from black mice in order to obtain pseudovirions containing only fragments of black-mouse DNA. The pseudovirions were purified and injected into pregnant white mice at different times in the gestation period. The progeny of the white mice were examined daily for about 1 month for the appearance of black hairs among the white hairs of the mice. The results were negative. No black hairs were detected in the progeny (Aposhian, unpublished).

2.1.7d. Other Biological Properties

Michel *et al.* (1967) have referred to experiments from which they conclude that polyoma pseudovirions do not induce cellular or viral DNA synthesis and are unable to direct the synthesis of progeny virus or viral hemagglutinin.

Skoog *et al.* (1970) have compared polyoma virions, pseudovirions, and empty capsids as to their activity in mouse embryo cells to induce expanded pools of the precursors of DNA and their activity to stimulate DNA synthesis. The limited stimulatory activity of the pseudovirions may be due to the measured contamination of the pseudovirions with infective particles as mentioned by the authors.

2.2. SV40 Pseudovirions

2.2.1. Introduction

The biology of polyoma virus and SV40 are very often discussed together since both are members of the papova group, both are used to study the mechanisms of viral carcinogenesis, and both appear to have similar replicative mechanisms. However, the pseudovirions of polyoma and SV40 will be discussed separately in order to allow continuity in the discussion of each.

There have been a number of extensive reviews of various aspects of SV40 biology (Winocour, 1969; Crawford, 1969*b*; Tooze, 1973; see also Chapt. 2 by N. P. Salzman and G. Khoury in Vol. 3 of this series).

A number of SV40-related particles have been reported. The infectious particles contain a genome which has a molecular weight of about 3.6×10^6 (Tai *et al.*, 1972). The DNA is circular, double-stranded, and supercoiled (Crawford and Black, 1964). Defective particles which appear to be spontaneous deletion mutants of SV40 have also been found. The repeated passage of SV40 at a high multiplicity of infection in African green monkey kidney (AGMK) cells results in about the same number of physical particles being found, but there is a marked decrease in the number of plaque-forming particles (Uchida *et al.*, 1966). Although the DNA of these defective particles is circular and supercoiled, the genome is 50–90% the size of the genome of infectious SV40 particles (Yoshiike, 1968). Such particles may be classified eventually as defective interfering viruses (Huang, 1973). Another type of defective SV40 particle has a partially deleted and partially substituted genome. Part of the SV40 genome has been deleted and replaced by sequences of host DNA covalently linked to the incomplete SV40 genome (Lavi and Winocour, 1972; Lavi *et al.*, 1973; Rozenblatt *et al.*, 1973). Whether the production of these particles containing partially deleted, partially substituted genomes will be affected by the different types of host cells in which the SV40 may be propagated, as is true for SV40 pseudovirus production, is not known as yet. Significant amounts of pseudovirions are found in SV40 preparations produced in AGMK cells and Vero monkey cells (Levine and Teresky, 1970; Trilling and Axelrod, 1970), but, as will be discussed below, the pseudovirions are not found after infection of some other types of monkey cells. Empty SV40 particles devoid of DNA are also found in impure preparations of the virus. Sokol *et al.* (1974) have reported the presence of a heavy form (ρ = 1.35 g/ml) and light form (ρ = 1.325 g/ml) of SV40 virions. It is not clear whether the heavy particles contain two double-stranded, circular DNA molecules of identical molecular size and the light particles contain only one, whether the light fraction contains aggregates of heavy virus and empty capsids, or whether some other explanation is possible for these unusual observations.

SV40 infection of monkey CV-1 cells and African green monkey cells induces the synthesis of cellular DNA (Gershon *et al.*, 1966; Hatanaka and Dulbecco, 1966; Kit *et al.*, 1967; Ritzi and Levine, 1970). A critical evaluation of experiments dealing with the induction of host DNA synthesis in SV40-infected cells has been made in a review by Winocour (1969).

2.2.2. Preparation and Purification

The preparation and identification of SV40 pseudovirions are possible because by radioactive labeling of the host-cell DNA before infection, followed by a thorough chase, only the DNA of the pseudovirions, and not the DNA of the infectious or defective virions, will be radioactive.

SV40 pseudovirions have been prepared by Levine and Teresky (1970) using primary AGMK cells. The cells, before infection, were labeled with ^{3}H-thymidine for two days and the radioactive thymidine was chased by adding unlabeled thymidine for two days. After the chase, the cells were infected with SV40 (large-plaque variant) using a multiplicity of infection of 25–100 PFU per cell. Four days after infection the cells were harvested and disrupted by sonication. Treatment with sodium deoxycholate, differential centrifugation, and RNase treatment were used for harvesting and preliminary purification of the virions. Final purification was performed using two cycles of equilibrium centrifugation in CsCl having a density of 1.32 g/ml.

Trilling and Axelrod (1970) have used Vero cells, a continuous line of green monkey cells, for the production of SV40 pseudovirions. The cells, labeled with ^{32}P-orthophosphate prior to infection, were infected with the small-plaque variant of SV40 at a multiplicity of infection of

5–50. The cells and culture medium were harvested 96 hr post-infection. The virus was purified by zone sedimentation onto a CsCl cushion and three cycles of equilibrium centrifugation in CsCl. The upper band of virus material, having a density of 1.341 g/ml, was collected, dialyzed, and treated with pancreatic deoxyribonuclease and pancreatic ribonuclease. Some of the properties of pseudovirions prepared and purified in this manner are summarized in Table 10. The upper band (Table 10), used as pseudovirions in their experiments, contained 38% pseudovirions. The lower band of the CsCl gradient, from which SV40 virions are obtained, contained 17% pseudovirions.

As yet, preparations of highly purified SV40 pseudovirions have not been prepared equivalent in purity to polyoma pseudovirions (1 polyoma particle per 10^6 pseudovirion particles), as reported by Osterman *et al.* (1970) and Qasba and Aposhian (1971). However, since the same techniques are involved, it should be possible to obtain such highly purified SV40 pseudovirions if desired.

2.2.3. Characterization

Levine and Teresky (1970) have analyzed the SV40 particles produced in primary AGMK cells. Virus was purified from cells that were labeled with ^{3}H-thymidine before infection. The purified virus was disrupted with 0.6% SDS and the DNA was analyzed in a neutral 5–20% sucrose gradient to which SV40 type I ^{32}P-DNA was added as a marker. The ^{3}H-DNA extracted from the particles sedimented between 11 S and 15 S, having an average sedimentation value of 14 S. The 14 S DNA was extracted from the particles and hybridized with AGMK DNA, which had been immobilized on a filter. The efficiency of hybridization was comparable to that found when AGMK ^{3}H-DNA is hybridized. The efficiency of the hybridization of the 14 S DNA with SV40 DNA was very low. Thus, the DNA in the SV40 pseudovirions is derived from cellular DNA and not from SV40 DNA. When particles produced in AGMK cells, labeled with ^{3}H-thymidine before infection, were examined in alkaline sucrose gradients, it was found that extensive single-stranded breaks are not present in the pseudoviral DNA.

Similar experiments have been performed with SV40 pseudovirions produced in Vero cells (Grady *et al.*, 1970). The extracted pseudoviral DNA was estimated to be about 15% smaller than SV40 DNA in molecular weight. However, it should be kept in mind that the size of pseudoviral DNA is dependent on the density of the SV40 pseudovirus particles from which the DNA is extracted.

TABLE 10

Characteristics of Upper- and Lower-Band SV40[a]

				Specific activities cpm/mg virus							
		Absorbance ratios		Post-infection labeled		Pre-infection labeled	Specific infectivities		Virus, %		
Band	Density, g/ml	260/242	260/250	0.25 μCi/ml	1.65 μCi/ml	1.0 μCi/ml	PFU/mg total virus	PFU/μg DNA type I	Pseudo-	Mature	Defective
Upper	1.322	1.09	1.40	165,000	848,000	446,000	1.44×10^{11}	1.5×10^{5}	38.2	28.1	33.7
Lower	1.332	1.08	1.37	150,000	836,000	219,000	3.43×10^{11}	3.3×10^{5}	17.3	82.7	—

[a] Axelrod and Trilling (1972).

Grady *et al.* (1970) have asked whether the fragments of monkey DNA found in SV40 pseudovirions represent random fragments of monkey DNA or whether the process by which the monkey DNA is fragmented is discriminatory so that an enrichment for specific sequences of monkey DNA results. Animal cell DNA contains unique sequences that occur only once per genome and others that exist in multiple copies, called repeated sequences (Britten and Kohne, 1968). Grady *et al.* (1970) determined the rates of reassociation of DNA prepared from pseudovirions, Vero cells and SV40 virions. The rates of reassociation indicated that all of the repeated sequences of the monkey cell DNA are represented in the SV40 pseudovirions. In order to determine whether unique sequences were also present, they mixed a small amount of pseudoviral ^{32}P-DNA with Vero ^{33}P-DNA and a large quantity of unlabeled Vero DNA, sheared the mixture, removed the repeated sequences, and determined the rate of reassociation of both labels. No significant difference between the reassociation rates of pseudoviral and Vero DNA was observed. The experiments indicate that the fragments of host DNA in SV40 pseudovirions contain repeated and unique sequences in the same proportions as found in the host DNA. They concluded that no one portion of the host genome is incorporated preferentially into SV40 pseudovirions.

Evidence that the protein coats within which the host DNA fragments are encapsidated are SV40 coats has been presented by Levine and Teresky (1970). Use was made of rabbit anti-SV40 antiserum and sheep anti-rabbit antiserum. Purified pseudovirions containing ^{3}H-DNA were incubated with the rabbit anti-SV40 serum followed by incubation with sheep anti-rabbit antiserum. The ensuing precipitate was collected by centrifugation and its radioactivity determined. The finding that 71% of the radioactivity in the purified pseudovirions was precipitated supports the conclusion that the pseudoviral DNA is within SV40 capsids. The purified pseudovirions sedimented at about 250 S, which is similar to the sedimentation rate of purified SV40 virus produced in BSC-1 cells. The BSC-1 cells do not produce pseudovirions, as will be discussed below.

2.2.4. Intracellular Events Related to Pseudovirion Production

Some of the intracellular events related to pseudovirus production in SV40-infected AGMK cells have been investigated by Ritzi and Levine (1973). The synthesis of cellular DNA begins 16–20 hr after infection, and 90% of the cellular DNA synthesis is completed by 45–50 hr after infection. The appearance of low-molecular-weight fragments of host DNA, as measured by the amount of radioactivity in the fractions soluble in SDS–1 M NaCl is first detected 45–60 hr after infection. This fragmentation of cellular DNA to a smaller size is called a very late event. The pseudovirions also are formed at a very late time after infection; a study of the kinetics of the formation of SV40 pseudovirions (Table 11) showed an eightfold increase between 36 to 60 hr and a fiftyfold increase between 60 and 96 hr post-infection.

By treating infected AGMK cells with cytosine arabinoside or 5-fluorodeoxyuridine at various times after infection, it was shown that cellular or viral DNA synthesis is required up to 40 hr after infection in order to obtain a level of cellular DNA fragmentation normally seen in SV40-infected cells. Also, protein synthesis is required up to 60 hr after infection to obtain the usual amount of cellular DNA fragments after SV40 infection.

Ritzi and Levine (1973), by using a temperature-sensitive mutant of SV40, also concluded that viral DNA synthesis is required for the efficient fragmentation of cellular DNA. At 41°C, cells infected with the SV40 mutant TSA-7 synthesized cellular DNA but did not synthesize viral DNA (Tegtmeyer and Ozer, 1971; Tegtmeyer, 1972). AGMK cells were labeled with radioactive thymidine before infection and then infected at 32° and 41°C with SV40 or SV40 TSA-7. The amount of fragmentation of cellular DNA at 96 hr or 144 hr after infection was determined by measuring the amount of radioactivity in the fraction

TABLE 11

The Kinetics of Formation of SV40 Pseudovirions in AGMK Cells[a]

Time after infection of virus harvest, hr	21 S DNA cpm	Increase	12–14 S DNA cpm	Increase
36	28,000		140	
		17×		8×
60	490,000		1,100	
		2.2×		50×
96	1,100,000		55,000	

[a] Cellular and virus DNA were labeled prior to and after SV40 infection of AGMK cells. At the times indicated, the cells were harvested, the virus purified, and viral (21 S) and cellular (12–14 S) DNA were extracted. The DNA was run on a neutral sucrose gradient and the levels of viral and cellular DNA in virions quantitated (Ritzi and Levine, 1973).

soluble in 1 M NaCl–SDS (Table 12). At 41°C, the TSA-7 mutant, which does not allow the synthesis of viral DNA but is permissive for host-cellular DNA synthesis, severely restricts the fragmentation of the host DNA.

Further examination of the cellular DNA after SV40 infection of AGMK cells indicates that 60–75% of the cellular DNA has extensive single-stranded breaks late in infection and that some 20–45% of the double-stranded cellular DNA is present in a low-molecular-weight form. It appears not only that the DNA fragments found in SV40 pseudovirions are representative of random pieces of host DNA, as shown by Grady *et al.* (1970), but the low-molecular-weight host DNA in the infected cell, which may be encapsidated eventually into pseudoviral particles, also represents random fragments of host DNA (Ritzi and Levine, 1973).

2.2.5. Influence of the Host Cell

A number of different types of monkey cells have been used for the production of SV40 virus. They include primary AGMK cells, as well as the continuous cell lines CV-1, BSC-1, and Vero. Levine and Teresky (1970) have made a most interesting observation concerning the influence of the host cell on the production of SV40 pseudovirions. AGMK, CV-1, or BSC-1 cells were labeled with ^{3}H-thymidine before infection. After infection with SV40, the particles produced in each cell type were purified and the DNA was extracted. Sedimentation of each

TABLE 12

The Effect of the TSA-7 Mutation on the Fragmentation of Cellular DNA after SV40 Infection[a]

Virus	Temperature, °C	Cellular DNA in the fraction soluble in 1 M NaCl–SDS, %
Mock infected	32	0.8
SV40 WT	32	13
TSA-7	32	15
Mock infected	41	1.8
SV40 WT	41	18
TSA-7	41	3.3

[a] At 32°C, the cells were harvested at 144 hr after infection. At 41°C, the cells were harvested at 96 hr after infection (Ritzi and Levine, 1973).

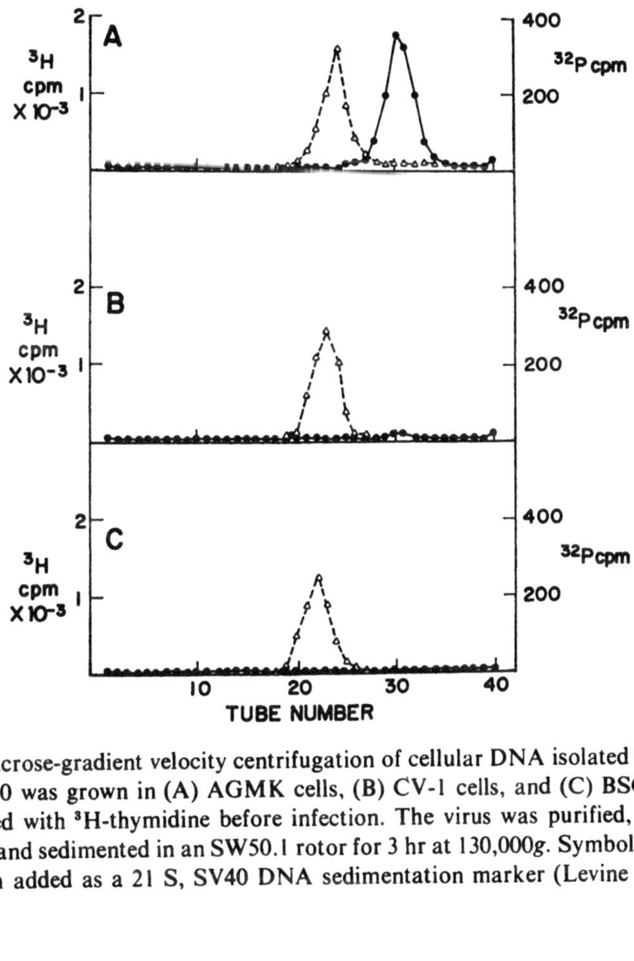

Fig. 9. Sucrose-gradient velocity centrifugation of cellular DNA isolated from purified SV40. SV40 was grown in (A) AGMK cells, (B) CV-1 cells, and (C) BSC-1 cells that were labeled with ^{3}H-thymidine before infection. The virus was purified, broken open with SDS, and sedimented in an SW50.1 rotor for 3 hr at 130,000*g*. Symbols: ●, ^{3}H cpm; Δ, ^{32}P cpm added as a 21 S, SV40 DNA sedimentation marker (Levine and Teresky, 1970).

of the purified DNA preparations in neutral sucrose gradients indicated that only the particles produced in the AGMK cells contained a component (Fig. 9) that sedimented at about 14 S. DNA–DNA hybridization experiments showed that the 14 S DNA hybridized with AGMK DNA with an efficiency comparable to the hybridization of radioactive AGMK DNA with nonradioactive AGMK DNA. However, purified virions produced in CV-1 cells, contained little if any 14 S host DNA fragments. Purified virus particles produced in BSC-1 cells contained no 14 S DNA. Therefore, Levine and Teresky (1970) concluded that SV40 infection produces pseudovirions in AGMK cells, but very little, if any, in CV-1 cells, and none in BSC-1 cells. The host-cell influence in the production of pseudovirions is much more clear-cut in the SV40 system (Levine and Teresky, 1970) than in the polyoma system (Yelton and Aposhian, 1973).

Ritzi and Levine (1970) have compared some of the intracellular events that occur in AGMK, CV-1, and BSC-1 cells during lytic infection with SV40 (Table 13). In each of these cell types they examined the kinetics of virus-induced cellular DNA synthesis and the conversion of high-molecular-weight cellular DNA to low-molecular-weight cellular DNA fragments. By pulse labeling the infected cells with ^{3}H-thymidine for 1 hr at different times post-infection and then analyzing the radioactivity in the SDS–1 M NaCl precipitate obtained from the cells, it was found that the kinetics of virus-induced cellular DNA synthesis in AGMK and CV-1 cells were very similar. However, little or no induction of cellular DNA synthesis was detected in the BSC-1 cells after infection. Furthermore, chemical determination, using the diphenylamine test, showed that the amount of cellular DNA increased about twofold after infection of AGMK and CV-1 cells, but, there was only a 1.1-fold increase in cellular DNA after infection of BSC-1 cells.

In AGMK cells infected with SV40, there is a 13–23% increase in the amount of low-molecular-weight cellular DNA produced from large-molecular-weight cellular DNA. The increase was ±1–2% for infected CV-1 cells and was absent in the infected BSC-1 cells. Ritzi and Levine (1970) have concluded that pseudovirions are produced in AGMK cells because the cells have the capacity for both the induction of cellular DNA synthesis and the excision of small fragments of DNA. Although the induction of cellular DNA synthesis occurs in CV-1 cells, the excision mechanism appears to be absent. On the other hand, BSC-1 cells are unable to induce cellular DNA or excise DNA fragments. These appear to be the reasons why CV-1 and BSC-1 cells are unable to produce SV40 pseudovirions. It should be kept in mind that although cellular DNA synthesis is not induced after SV40 infection of the strain

TABLE 13

Summary of Data for AGMK, CV-1, and BSC-1 Cells[a]

	Time of appearance (hr) of virus-specific:		Cellular DNA		
Cell line	DNA	Infectious SV40	Induction	Excision	Pseudovirions
AGMK	15–20	20–24	+	+ (13–23%)	+
CV-1	15–20	20–24	+	± (1–2%)	–
BSC-1	30–38	36–45	–	–	–

[a] Ritzi and Levine (1970).

TABLE 14

Fragmentation of Large-Molecular-Weight Cellular DNA after Infection with SV40[a]

Cell line	Expt. No.	Percentage of total cellular DNA in 1 M NaCl–SDS soluble fraction at			
		95–96 hr		125 hr	
		+SV40	−SV40	+SV40	−SV40
AGMK	1	15.1	2.2	—	—
	2	27.5	4.7	—	—
CV-1	1	5.3	2.7	—	—
	2	4.8	3.4	—	—
BSC-1	1	3.1	2.9	—	—
	2	2.8	2.7	3.8	3.2

[a] These data demonstrate the percentage of the total cellular DNA that is found as small-molecular-weight fragments in AGMK, CV-1, or BSC-1 cells at 96 or 125 hr after infection (Ritzi and Levine, 1970).

of BSC-1 cells used by Levine and his associates, cellular DNA synthesis is induced after SV40 infection in another BSC-1 cell line (Sheppard *et al.*, 1971).

As Ritzi and Levine (1970) point out, the different responses of these three monkey cell lines (Table 14) suggest the influence of host-cellular functions in the induction of host-cell DNA synthesis and in the excision of host DNA fragments for pseudovirus production.

It is hoped that the SV40–AGMK system might be used to find the endonuclease that cleaves large-molecular-weight cellular DNA to pseudoviral DNA size. Experiments to determine the presence of a specific endonuclease in the uninfected or infected AGMK cells and the lack of such proteins in the CV-1 or BSC-1 system might lead to the identity of such an enzyme. It would be of interest, also, to determine whether the SV40-associated endonuclease (Kaplan *et al.*, 1972; Kidwell *et al.*, 1972) is associated with SV40 virions harvested from AGMK, CV-1, and BSC-1 cells. Also, the influence of this SV40-associated endonuclease on monkey DNA would be of interest and this information might help to unravel the enzymatic mechanism for the production of pseudoviral DNA.

It appears that the first reports dealing with the SV40-associated endonuclease involved a contaminating endonuclease derived from the serum of the medium (Ozer, personal communication). Until definitive results are obtained, this enzyme should be considered a contaminant and of questionable biological significance.

2.2.6. Biological Properties

Grady *et al.* (1970) have used SV40 pseudovirions which contain fragments of monkey cell (Vero) ^{32}P-labeled DNA to infect monolayers of C-57 mouse embryo cells. Mouse cells were used since they do not undergo a lytic response to SV40 virions, which were present in substantial amounts in the pseudovirus preparation. About 6–15% of the pseudoviral radioactivity was taken up by the mouse cells (Table 15). The nuclear fraction and cytoplasmic fraction of such infected cells were also prepared. Lysis of the nuclei and sedimentation through alkaline sucrose gradients showed the presence of SV40 pseudovirions in the nuclear fraction of infected cells. This type of experiment does not differentiate between the presence of SV40 pseudovirions bound to the outer surface of the nuclear membrane and those within the nucleus, but only shows that they are present in the nuclear fraction. As yet, there are no reports as to whether SV40 pseudovirions are uncoated in such mouse cells.

A. J. Levine (personal communication) has attempted transductions with SV40 pseudovirions using as recipients C_2F-3T3-TK$^-$ cells. The cells were infected with pseudovirions at a multiplicity of about 100–1000 particles per cell. The pseudovirions were prepared in primary or secondary AGMK cells. The light-density fraction (in CsCl) of purified SV40 virions was used for infection. About 40–60% of the particles were pseudovirions, based on sedimentation of the DNA in sucrose gradients. Cells or colonies were tested for the transfer of the TK gene by measuring ^{3}H-thymidine incorporation as judged by acid-precipitable radioactivity or by autoradiography. All attempts were negative.

2.3. Other Animal Virus Systems Studied for Pseudovirions

There has been only one other animal virus system in which a search for pseudovirions has been reported; zur Hausen (1969) has reported the failure to detect incorporation of host-cell DNA into virions of adenovirus type 12.

However, an interesting group of papovaviruses has been found in humans. The JC virus and PML-2 virus have been isolated by Padgett

TABLE 15

Incorporation of SV40 Pseudovirus by Mouse Cells[a]

	Time after infection, hr	Cell uptake, %	Distribution in cells: Nonnuclear, %	Distribution in cells: Nuclear, %
Stationary culture	18	14.6	93.0	7.0
	42	12.9	92.2	7.8
Dividing culture	66	6.1	90.6	9.4

[a] Grady *et al.* (1970).

et al. (1971) and Weiner *et al.* (1972), respectively, from patients with progressive multifocal leucoencephalopathy. Gardner *et al.* (1971) have isolated an apparently different papovavirus, called BK, from a renal allograft recipient. The JC and PML-2 viruses have been shown to be serologically related to SV40. Since these viruses seem to be very similar to SV40 and since they have been propagated, also, in cultured cells, it should not be surprising to find eventually that JC, PML-2, or BK populations have pseudovirions containing fragments of human DNA. As yet, however, such observations have not been reported.

3. PLANT CELL PSEUDOVIRIONS (TOBACCO MOSAIC PSEUDOVIRIONS)

3.1. Separation

Siegel (1971) has reported the occurrence of particles, containing host RNA, in preparations of tobacco mosaic virus (TMV) and has called these particles TMV pseudovirions. A partial separation of the pseudovirions from TMV particles was accomplished by electrophoresis of radioactive TMV preparations in 0.5% agarose gels. After the electrophoresis, the gel was scanned for optical density to detect the position of the major portion of viral material. This was followed by slicing the gel and extracting the RNA from individual slices. Then, the RNA from each fraction was incubated with nitrocellulose membrane filters containing tobacco chloroplast DNA. When the counts per minute that hybridized to the DNA were plotted versus the position of the slice in the gel, it was found that the hybridizable material had moved farther in the gel than the major portion of the TMV, as judged by the optical density peak and the ^{3}H-RNA peak. Since the pseudovirions moved farther into the gel than did the TMV particles, it appears that the pseudovirions are smaller, on the average, than the TMV particles. Unfortunately, no experiments using velocity sedimentation as a means of separating these particles have been reported.

3.2. Characterization

3.2.1. Identification of Pseudoviral RNA

Evidence for host RNA being present in these pseudovirus particles is based on DNA–RNA hybridization experiments (Siegel, 1971). The pseudovirus content of various TMV strains has been found to differ. The U2 strain contains about 2–2.5% pseudovirions. The U1, Holmes' ribgrass, and TS38 strains of TMV contain smaller amounts. The pseudoviral RNA appears to be unique and does not represent a random selection of host RNA. Again, the evidence is based on DNA–RNA hybridizations and indicates that pseudoviral RNA appears to have been transcribed, primarily, from chloroplast DNA and very little, if any, consists of ribosomal RNA. U2 pseudovirus RNA is complementary to about 20% of tobacco chloroplast DNA and 0.4% of tobacco nuclear DNA. The molecular weight of pseudoviral RNA has not been determined although, based on electrophoretic mobility in gels, it appears to be smaller than TMV RNA.

3.2.2. Identification of Proteins

Unfortunately, there have been no reports on the properties of the protein coats of the pseudovirions that have been found in TMV preparations. It is possible that the coats might be unrelated to TMV coats. For example, the orphan pseudovirions that are induced in *Bacillus subtilis* by physical or chemical agents can also be induced by infecting *B. subtilis* with phage SP3 (see Sect. 5). There is no question that the particles found in TMV preparations are pseudovirions. The question, however, remains whether they are pseudovirions of TMV or whether they are in the orphan pseudovirus classification. The major reason for calling these pseudovirions, at present "TMV pseudovirions" is that they contain tobacco RNA and are found in TMV preparations.

4. BACTERIAL CELL PSEUDOVIRIONS (GENERALIZED TRANSDUCING PHAGES)

Although the term "pseudovirus" was used originally in 1967 by Michel *et al.* (1967) to describe particles consisting of fragments of mouse DNA encapsidated in polyoma coats, similar particles consisting of phage coats encapsidating fragments of bacterial DNA, but virtually free of viral DNA, had been known for about 14 years. Some of these phagelike particles had been shown to be capable of transferring bacterial genes from one cell to another (Zinder and Lederberg, 1952). This type of gene transfer has been called phage-mediated transduction or, more simply, transduction. The particles capable of transduction have been called transducing particles to distinguish them from normal phage particles. Externally, transducing particles are indistinguishable from normal phage particles as far as their size, shape, adsorption characteristics, serological properties, etc. (Zinder, 1953; Stocker, 1958; Campbell, 1964).

Two types of transduction have been observed, namely, generalized and specialized transduction. Generalized transducing particles produced in phage-infected bacterial cells fit the definition of pseudovirions. Specialized transducing particles do not. These bacterial-cell pseudovirions are the only pseudovirions that have been demonstrated, experimentally, to transfer genetic information from one cell to another. Examples of generalized transduction are P22 in *Salmonella typhimurium* (Zinder and Lederberg, 1952) and P1 in *Escherichia coli* (Lennox, 1955). Almost any genetic marker of a donor strain can be transferred by generalized transducing particles. On the other hand, phage λ in *E.coli* K12 (Morse *et al.*, 1956) is able to transduce specialized markers only, such as those of the *gal* (galactose) loci.

A review summarizing the evidence for various models to explain generalized and specialized transduction in bacterial systems is available, as is a listing of previous reviews (Ozeki and Ikeda, 1968).

Ozeki and Ikeda (1968) have proposed the "wrapping choice model" for the formation of generalized transducing particles. The model is essentially that proposed by Zinder (1953, 1955) and by Stocker (1958). In this model, the coats of phages, capable of forming generalized transducing particles, are believed to be able to wrap bacterial DNA fragments of a specific size. The fragments of bacterial DNA are believed to be wrapped accidentally in place of the phage DNA. Inherent in the model is a mechanism for cutting the bacterial DNA to a size suitable for wrapping.

Phages known to be capable of carrying out generalized transductions may also perform specialized transductions. However, phages known to perform specialized transductions are not necessarily able to carry out generalized transduction. Ozeki and Ikeda (1968) suggest that this is because some specialized transducing particles are produced by the "wrapping choice model," although this type of particle is more commonly produced by the "hybrid formation mechanism." The "hybrid formation mechanism" is based on prophage integration. For example, this model considers the basic mechanism of the specialized transduction of *gal* by phage λ to be the replacement of part of the phage chromosome by a portion of the bacterial chromosome. Specialized transduction is restricted to the region of the bacterial chromosome adjacent to the prophage attachment site.

Although it is not the intent of this chapter to review the extensive literature of the generalized transducing phages, two elegant papers have appeared that might serve as models for a study of animal cell pseudovirions once a greater number of genetic markers are available. These papers by Ebel-Tsipis *et al.* (1972*a,b*) deal with the molecular origin and integration of the DNA of P22 transducing particles.

Ebel-Tsipis *et al.* (1972*a*) prepared generalized transducing particles of phage P22 using media containing ^{15}N, ^{2}H, and ^{14}C. Three types of lysates were prepared: (1) a light–light lysate in which cells were grown in light medium before infection and maintained in light medium after infection, (2) a heavy–heavy lysate in which heavy medium was used both before and after infection, and (3) a heavy–light lipate in which cells were grown in heavy medium and then transferred to light medium after phage adsorption. The resulting phage and their DNA were purified and their buoyant densities and transducing activities determined. The transducing particles were found to be very similar in their physical properties to phage particles. As judged by zone sedimentation, the particles are of the same size. The DNA of the two types of particles was shown to have the same molecular weight. Ikeda and Tomizawa (1965) reached similar conclusions about the properties of the coliphage P1.

Ebel-Tsipis *et al.* (1972*a*) also concluded that P22 transducing particles contain primarily bacterial DNA which was synthesized before infection and have little or no phage DNA in them. As is the case for all pseudovirions, the specific mechanism for the encapsidation of host DNA by P22 particles is not understood. However, it does not appear to require the general recombination functions that are known, at present, for *S. typhimurium* infected with P22.

The mechanism of the integration of the DNA of the P22 transducing particle into the recipient cell DNA has also received recent study (Ebel-Tsipis *et al.*, 1972*b*). Using purified transducing particles labeled with ^{3}H, ^{2}H, ^{15}N, and ^{13}C, as well as recipient cells labeled with ^{32}P, these authors were able to draw a number of conclusions. About 12–15% of the injected DNA of the transducing particles become associated with the recipient-cell DNA; 90–95% of the transducing DNA which eventually becomes associated with the DNA of the recipient cell is covalently linked with that DNA. By 20 min after transduction, double-stranded fragments of transducing DNA are found integrated into the DNA of the recipient cell. However, only 20–30% of the donor DNA which becomes associated with the recipient-cell DNA is integrated as large fragments. This means that only 2–5% of the total transducing DNA taken up by the bacteria is integrated as large fragments. The rest of the DNA from the transducing particles appears to be broken down to precursors of DNA synthesis and reutilized. The molecular weight of most of these large integrated fragments is $2–4 \times 10^{6}$, which is much less than 27×10^{6}, the size of the DNA originally in the transducing particles.

Some of the other generalized transducing particles that have been studied by various groups are the P1–*E. coli* system (Ikeda and Tomizawa, 1965*a,b*); the SP10–*B. subtilis* system (Okubo *et al.*, 1963); and the PBS1-*B. subtilis* system (Yamagishi and Takahashi, 1968).

Much can be learned about pseudovirions from other sources by using as prototypes the many experiments designed to elucidate the molecular properties and mechanisms of pseudovirions of bacterial cells.

5. ORPHAN PSEUDOVIRIONS

5.1. From *Bacillus subtilis*

Treatment of most strains of *B. subtilis* with any of a variety of inducing agents results in the lysis of the cells and the appearance of phagelike particles in the lysate (Seaman *et al.*, 1964; Haas and Yoshikawa, 1969*a*). These particles have been found to contain fragments of *B. subtilis* DNA and appear to be devoid of the DNA of any known temperate or virulent phage. Since the particles are unable to reproduce when added to cultures of all the strains of *B. subtilis* tested, they have been called defective bacteriophages and protophages

(Seaman *et al.*, 1964; Haas and Yoshikawa, 1969*a*). If the original viral genome containing the genes that direct the synthesis of the protein coats of these particles were known, these particles would fit the definition of pseudoviruses since they have fragments of host DNA. However, since the original genome is not known, these phagelike particles will be referred to as orphan pseudoviruses in this chapter. The present discussion will be limited to those orphan pseudovirions induced in *B. subtilis* that have been studied and characterized most extensively. The occurrence of these and similar particles from other bacteria has been listed and reviewed by Garro and Marmur (1970). The reader is referred to their review for references to other orphan pseudoviruses that have been less well characterized.

5.1.1. Induction and Preparation

Although small amounts of these orphan pseudovirions are produced in cultures of *B. subtilis* that have not been purposely induced, induction by chemical or physical agents is the most efficient method of obtaining them (Haas and Yoshikawa, 1969*a*). A number of orphan pseudovirions have been induced by treating *B. subtilis* with these agents. Of these, PBSX (Seaman *et al.*, 1964), α (Eiserling, 1964), μ (Ionesco *et al.*, 1964), GA2 (Bradley, 1965), PBSH (Haas and Yoshikawa, 1969*a*), and ϕ3610 (Stickler *et al.*, 1965) are very similar particles and may be identical. They are different from PBSY and PBSZ, as will be shown below. PBSX, PBSY, PBSZ, and PBSH are induced most commonly by treating appropriate strains of *B. subtilis* with mitomycin C (Seaman *et al.*, 1964; Haas and Yoshikawa, 1969*a*). PBSH has also been induced by thymineless germination, germination with 5-bromodeoxyuridine, and by ultraviolet light, but these inducing agents produce fewer PBSH particles than when the cells are treated with mitomycin (Haas and Yoshikawa, 1969*a*). All of these induction methods produce PBSH particles that are similar in appearance when viewed using the electron microscope. Boice (personal communication) has been able to induce PBSX particles by treating spores of *B. subtilis* 168, *thy*$^-$ *ind*$^-$ with bromouracil. Phage ϕ3610 has been induced by treating *B. subtilis* 3610 with hydrogen peroxide (Stickler *et al.*, 1965).

Siegel and Marmur (1969) have shown that PBSX can be induced in the absence of chemical agents. They made use of *B. subtilis* MB500, a temperature-sensitive mutant of *B. subtilis* 168. The mutant forms colonies at 34°C but is unable to do so at 48°C. When MB500 is grown in broth at 48°C, the culture begins to lyse after 90 min and the lysate contains PBSX particles, as shown by a number of criteria. This temperature-sensitive induction of PBSX is specific for MB500. When other temperature-sensitive strains blocked in other functions are used, PBSX induction does not occur.

Induction of PBSH is limited to a narrow interval during early exponential growth (Haas and Yoshikawa, 1969*a*). Under optimal conditions, 400 PBSH particles are liberated from each cell of strain 168 LTT. In the case of PBSX, approximately 125 particles are liberated per cell (Okamoto *et al.*, 1968*a*), but the optimal conditions for PBSX production may not have been found.

Seaman *et al.* (1964) tested several species and strains of *Bacillus* for inducibility with mitomycin. The kinetics of induction by mitomycin and the release of PBSX were followed using viable cell count, optical density change, and killing activity of PBSX particles as assayed on a cured strain. All of the *B. subtilis* strains tested and most of the closely related species were found to be lysogenic and to release PBSX-like particles.

Haas and Yoshikawa (1969*a*), found, however, that not all strains of *B. subtilis* which are lysed by treatment with mitomycin C will necessarily produce orphan pseudovirions (Fig. 10). Strain 168 LTT lyses very abruptly within 90 min after resuspension, and phage are produced. Strain 168 WT produces phage efficiently even though part of the culture does not lyse. However, although a slow lysis of strain 168 LTT CA1 occurs, phage are not produced. Finally, strain W23 LH shows neither lysis nor detectable phage production after treatment with mitomycin C.

The induction of PBSX is affected by chloramphenicol (Okamoto *et al.*, 1968*b*). Neither breakdown of the bacterial DNA to the specific size of pseudoviral DNA nor the production of detectable particles is observed in the culture if chloramphenicol is added at 0, 20, or 30 min after induction by mitomycin. When chloramphenicol is added 43 min after induction, a small number of PBSX particles are found. The inhibitor of DNA synthesis, 5-fluorodeoxyuridine, when added with mitomycin C, appears to inhibit the rate of PBSX production and to decrease the final yield of phage (Okamoto *et al.*, 1968*b*).

To obtain these *B. subtilis* orphan pseudovirions (Seaman *et al.*, 1964; Okamoto *et al.*, 1968*a*; Haas and Yoshikawa, 1969*a*), an appropriate strain of *B. subtilis* is grown, usually to a titer of about $1–3 \times 10^7$ cells per ml, and treated for a short period of time with mitomycin C. After the unbound mitomycin is removed by either centrifuging or filtering the culture, the treated bacteria are resuspended. Lysis may occur as early as 90 min or may take a longer period of time, depending

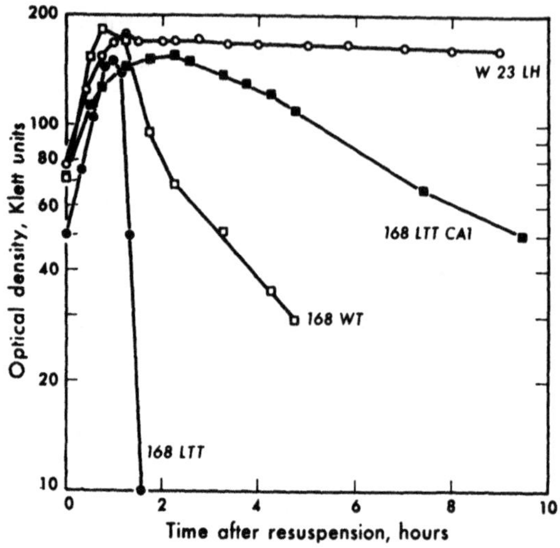

Fig. 10. Growth and lysis of four *B. subtilis* strains. Cultures were induced with 4 μg of mitomycin C per ml for 15 min before resuspension in the same medium without mitomycin (Haas and Yoshikawa, 1969*a*).

on the strain of *B. subtilis* used. The resulting particles are purified by the common methods of virus purification, e.g., differential or isopycnic centrifugation.

5.1.2. Characterization

5.1.2a. Physicochemical Properties

These particles have been characterized by electron microscopy, by their activity in killing various strains of *B. subtilis*, and by examining their DNA. The orphan pseudovirions PBSX and PBSH are considered to be similar or identical (Garro and Marmur, 1970). The physical dimensions of the particles, within experimental error, are similar (Table 16), as are their densities in CsCl. Their appearance is similar to many virulent and temperate phage. However, the diameter of the heads of PBSX and PBSH is unusually small relative to the diameter of their tails.

5.1.2b. Killing Range

The killing activity of these particles on different strains of *Bacillus* has been studied and used to characterize some of these orphan pseudovirions (Okamoto *et al.*, 1968*a*). The hosts in which some of these phagelike particles are produced and the killing activities of the particles are listed in Table 17. The particles produced by a particular strain will often kill other strains but the strain from which the particle is produced is immune to its killing action. This immunity appears to be due to the inability of the pseudovirus to be adsorbed to the strain of *B. subtilis* from which it was derived (Okamoto *et al.*, 1968*a*). Although these orphan pseudovirions have killing activity, they are unable to inject their DNA or form plaques (Okamoto *et al.*, 1968*a*).

One of the features that distinguishes PBSX from bacteriocins is the inability of PBSX to be adsorbed to strains that are immune to it. In contrast, bacteriocins are adsorbed to cells immune to their actions. In addition, bacteriocins are bacterial proteins coded for by episomes. The genes controlling the synthesis of PBSX tails are located on the bacterial chromosome, as will be discussed below.

TABLE 16

Three Very Similar Orphan Pseudovirions from *Bacillus* Cells

	PBSX	PBSH	PBLB
Head diameter	41 nm	41 nm	41 nm
Tail length	192 nm	200 nm	—
S°_{20}	160	248	160
Density	1.375 g/ml	1.375 g/ml	1.372 g/ml
Particle wt.	—	68×10^6	—
DNA			
S°_{20}	22	25.35	22
Mol. wt.	8.35×10^6	12×10^6	—
Density	1.703 g/ml	1.705 g/ml	1.706 g/ml
G + C, %	43	43	—
T_m	87.5	86	—
% DNA	21	16.5	—
Host	*B. subtilis 168*	*B. subtilis 168*	*B. licheniformis* NRS 243
Reference	Okamoto *et al.* (1968*a*)	Haas and Yoshikawa (1969*a*)	Huang and Marmur (1970*b*)

TABLE 17

Killing Range of the Defective Phages of *B. subtilis*[a]

Defective phage	Produced by	Killing activity indicator strains		
		168	S31	W23
PBSX	168	−	+	+
PBSY	S31	+	−	+
PBSZ	W23	+	+	−

[a] Huang and Marmur (1970*a*).

5.1.2c. Identification and Origin of the DNA

The DNA of these orphan pseudovirions consists of linear, double-stranded fragments of host-cell DNA. Evidence for this is based on physicochemical data, DNA–DNA molecular hybridizations, transformation of *B. subtilis* chromosomal markers using purified orphan pseudoviral DNA, and, finally, kinetic experiments concerning the conversion of bacterial DNA to pseudoviral DNA.

The DNA extracted from purified particles of either PBSX or PBSH has a density in CsCl identical to *B. subtilis* DNA (Okamoto *et al.*, 1968*a*; Haas and Yoshikawa, 1969*a*). PBSX DNA sediments at 22 S in neutral sucrose (Okamoto *et al.*, 1968*a*). Haas and Yoshikawa (1969*a*) found that PBSH DNA sediments at 25 S; they discuss a number of factors that may influence the calculation of the sedimentation constants.

One of the strongest pieces of evidence that pseudoviral DNA consists of fragments of *B. subtilis* DNA is based on DNA–DNA hybridization studies. Okamoto *et al.* (1968*a*) labeled the bacterial DNA with [^{3}H]-thymidine before induction with mitomycin C. After induction, the PBSX particles were purified and their DNA was extracted. The purified DNA contained [^{3}H]-thymidine and was found to hybridize with *B. subtilis* DNA with an efficiency identical to the hybridization of *B. subtilis* DNA with itself. Hybridization competition experiments using PBSX DNA and *B. subtilis* DNA further supported the identity of PBSX DNA as fragments of *B. subtilis* DNA.

In addition, the same investigators showed that pseudoviral DNA was capable of the transformation of four unlinked markers of *B. subtilis*. The *ade, arg, leu,* and *met* markers were transferred with an efficiency equal to or in some cases greater than that normally found when *B. subtilis* DNA was used. Obviously, all these results cannot be attributed to a mechanism whereby pseudoviral DNA would be synthesized from the deoxynucleoside and deoxynucleotide breakdown products of host-cell DNA.

The analysis by Huang and Marmur (1970*a*) of the 5′ ends of the DNA from PBSX, PBSY, and PBSZ further indicates the different identities of these three orphan pseudovirions. The DNAs were extracted from each of these phage by phenol and their 5′ termini labeled with ^{32}P using T4-polynucleotide kinase. By treating the purified product of the kinase reaction with pancreatic DNase and snake venom phosphodiesterase, the orphan pseudoviral DNAs were hydrolyzed to nucleoside 5′-monophosphates. The distribution of the ^{32}P among the four deoxynucleoside monophosphates was determined by electrophoresis (Table 18). The results indicate that the 5′ ends of the DNA of two of these orphan pseudoviruses are clearly unique. The 5′-terminal nucleotides for both PBSX and PBSY are dGMP and dTMP. In the case of PBSZ, dAMP and dTMP are the predominant 5′-terminal nucleotides, although dCMP and dGMP were found in significant amounts. Sonicated *B. subtilis* DNA, the 5′-termini of which were labeled as a control, had all four nucleotides represented in the 5′-terminal position in amounts that would be expected according to its base composition. Huang and Marmur (1970*a*) further reported preliminary results using λ exonuclease and *E. coli* exonuclease I which indicate that PBSX does not possess significant single-stranded regions at either the 3′ or 5′ termini of the molecule.

As yet there has been no report of the identification or isolation of the enzymes responsible for the cleavage of *B. subtilis* DNA having a

TABLE 18

Identification of 5′-Terminal Nucleotides of PBSX, PBSY, and PBSZ DNA[a]

	Percent of total		
	PBSX DNA	PBSY DNA	PBSZ DNA
dCMP	7	0.5	19
dAMP	3	0.5	32
dGMP	43	44	13
dTMP	47	55	37

[a] Huang and Marmur (1970*a*).

molecular weight of about 1×10^9 to the pseudovirus fragments having a molecular weight of 8.4×10^6. As Huang and Marmur (1970*a*) have stated, the high efficiency of the process in which up to 50% of the host DNA can be packaged into phage particles following mitomycin C induction makes the *B. subtilis* orphan-pseudovirus system very amenable for the study of these enzymes. Perhaps, an almost insurmountable obstacle, at present, to finding the enzymes responsible for the production of orphan-pseudoviral DNA is the routine preparation of *B. subtilis* DNA with a sufficiently high molecular weight to act as a suitable substrate. Certainly, if the molecular weight of the product of the reaction is about 8.4×10^6, the molecular weight of a suitable substrate might be expected to be, at a minimum, about 16.8×10^6, and maximally probably 10^9. At present there do not seem to be reproducible methods for preparing such large-molecular-weight DNA of sufficient purity and in sufficient amounts for enzyme assays. The isolation of such enzymes which appear to yield fragments having dGMP or dTMP at their 5′ termini would be immensely helpful as they could be used as laboratory reagents for DNA sequence analysis as well as tools for understanding the mechanism by which orphan-pseudoviral DNA is produced from the large-molecular-weight host DNA.

5.1.2d. Origin of the Proteins

Garro *et al.* (1970) have demonstrated that the *xtl* marker, responsible for PBSX phage tail production, is located on the bacterial chromosome and linked to the *met c* marker. These experiments were based on transformation and PBS1-mediated transduction, as well as on the observation that the 44AO and BS10 mutant strains of *B. subtilis* 168, when induced with mitomycin, produce PBSX heads but fail to make phage tails. Whether the PBSX genome is scattered or localized on the *B. subtilis* chromosome is still unknown.

No information is available concerning the identity of the original genome which codes for the synthesis of the coats of orphan pseudovirions. There have been no reports as to whether the coats of PBSX have any serological cross-reactivity with any *B. subtilis* proteins or any components of the other well-known virulent or temperate phages of *B. subtilis*. Seaman *et al.* (1964) have reported that there is no serological cross-reactivity between PBSX and PBSZ lysates. However, Huang and Marmur (1970*b*) have cited preliminary experiments which indicate that antiserum against PBSX is also able to neutralize PBSZ and the *B. lichenformis* orphan pseudovirus PBLB.

5.1.3. Intracellular Events Related to Pseudovirus Synthesis

Some of the intracellular events associated with the synthesis of *B. subtilis* orphan pseudovirions have been examined and may be summarized as follows. Induction by mitomycin C may involve the synthesis of an immunity-destroying substance (Haas and Yoshikawa, 1969*a*) and/or a DNase that cleaves the host DNA to orphan-pseudoviral DNA of about 8×10^6 daltons. Also, genes located on the bacterial chromosome (it is not known whether these genes are part of the genome of a temperate virus) are derepressed and the synthesis of pseudovirus coat proteins begins. Perhaps, in accordance with the hypothesis that the head is full of DNA, these phage heads or an associated protein may be responsible for the cutting of host DNA to the size of 8×10^6 daltons once the virus head is packed with sufficient DNA.

PBSH appears to package fragments of host DNA that are located close to the origin of replication of the *B. subtilis* chromosome. Using the DNA prepared from pseudovirions, Haas and Yoshikawa (1969*b*) showed that many bacterial markers were transformed at a frequency identical to that found for *B. subtilis* DNA, except that the frequency of transformation of the *ade-16* marker was increased 30–100 times. The *ade-16* marker is located near the origin of the bacterial chromosome and the induction by mitomycin C may initiate multiple initiation points for synthesis of bacterial DNA. The properties of the *ade-16* marker in purified pseudoviral DNA have been examined by Haas and Yoshikawa (1969*c*). The marker was found to renature more rapidly and to a greater extent than any other marker in the pseudoviral DNA or in the bacterial DNA. Possible causes of this unusual behavior have been discussed by Haas and Yoshikawa (1969*b,c*). However, Siegel and Marmur (1969) have suggested that the greater transformation frequency of the *ade-16* marker might be related more to the mitomycin C treatment than to PBSX induction. Their suggestion is based on their finding that DNA prepared from temperature-induced PBSX does not exhibit a greater transformation frequency for the *ade-16* marker.

Okamoto *et al.* (1968*b*) have shown that after induction 50 S DNA of *B. subtilis* is cleaved to fragments which sediment at 22 S, the same size as pseudoviral DNA. No DNA molecules intermediate in size are detected. About 40% of the host DNA existing before induction is found later in the particles. In the mature pseudovirions, about 14% of the particle DNA is derived from host DNA synthesized prior to induction, and about 86% is derived from host DNA replicated after infection. About 25% of the preexisting host DNA was also degraded to a

TCA-soluble form. A 16 S DNA species has been found late after induction of PBSY (Okamoto *et al.*, 1968*b*). The significance of this 16 S fragment is not understood. However, Siegel and Marmur (1969) did not detect any 16 S DNA when the PBSX was induced in the temperature-sensitive strain MB500.

As yet, no evidence has been found for the replication of the discrete pieces of orphan pseudovirus DNA (Haas and Yoshikawa, 1969*b*). Rather, the DNA appears to arise from fragmentation of the host DNA.

Since pseudoviral DNA is very homogenous in size and has only a few, if any, single-stranded breaks, Okamoto *et al.* (1968*b*) argue against the possibility of a random breakage of host DNA and subsequent selection of only the DNA molecules of unique size for packaging. Rather, they favor the hypothesis that the host DNA is broken down to a unique size and then wrapped up into phage particles in line with the headful hypothesis proposed by Streisinger *et al.* (1967) for T4. Strong support for this is the almost exclusive presence of either dGMP or dTMP at the 5′ terminus of PBSX DNA (Huang and Marmur, 1970*a*).

5.1.4. Biological Properties and Defectiveness

Within 5 min after exposing *B. subtilis* W23 to PBSX particles, the synthesis of protein, DNA, and RNA are inhibited (Okamoto *et al.*, 1968*a*). Okamoto *et al.* offer one explanation for the defectiveness of PBSX particles. Using a Hershey–Chase type of experiment, they were able to demonstrate that PBSX particles cannot inject their DNA into the cells to which they adsorb. However, Haas and Yoshikawa (1969*b*) have concluded that PBSH is defective in DNA replication. Since the DNA does not appear to be injected into any cells to which it is adsorbed, the importance of the latter type of defectiveness is questionable. Although Haas and Yoshikawa (1969*b*) used PBSH particles to "transform" *B. subtilis*, these orphan pseudovirions do not form plaques on any organism yet tested. Their results may be explained by the presence of free DNA in their PBSH preparations.

5.2. From *Bacillus licheniformis*

Huang and Marmur (1970*b*) have demonstrated that another species of the *Bacillus* genus produces orphan pseudovirions. The treatment of *B. licheniformis* with mitomycin C produces two different defective phage particles, designated PBLA and PBLB. The former particle will not be classified as an orphan pseudovirion since it does not appear to contain host DNA sequences exclusively. On the other hand, PBLB fits the definition of an orphan pseudovirion and is very similar to PBSX obtained from *B. subtilis* (Table 16). Antiserum against PBSX neutralized PBLB. *B. lichenformis* DNA synthesized before and after induction is found in PBLB.

6. POSSIBLE BIOLOGICAL ROLES OF PSEUDOVIRIONS

Do pseudovirions have any important biological role or, after being produced in virus-infected cells, are they at a biologic and genetic dead end? The answers to these questions must be speculative since much still remains to be learned about these interesting particles. There is, of course, the temptation to consider pseudovirions as vectors for transferring genetic information from one cell to another. Certainly, as far as the pseudovirions of bacterial origin (the generalized transducing phages) are concerned, their role in transferring genetic information has been established and this transfer may have some limited evolutionary importance. In the case of the pseudovirions of animal cells, the transfer of DNA from one cell to another by these particles has been demonstrated. However, attempts to transfer specific genetic information by means of animal cell pseudovirions have so far been unsuccessful. Furthermore, animal cell pseudovirions have been shown to be produced in cultured cells only. No evidence exists as to whether they are found after virus infection of the whole animal. Therefore, these particles may be merely a laboratory curiosity.

However, another interesting possibility as to the role of these pseudovirions is that they may act as a means by which the cell decreases the severity of a polyoma or SV40 infection. Certainly, in the case of polyoma infection of mouse embryo cells, more pseudoviral DNA than polyoma DNA is produced. These two types of DNA appear to compete in the virus assembly process so that more pseudovirions than polyoma virions are produced when mouse embryo cells are infected. The modulation of a number of different virus infections by defective interfering particles has been reviewed by Huang (1973). Although pseudovirions do not fit the definition by Huang (1973) of defective interfering particles, perhaps they are capable of also interfering with the reproduction of nondefective virions.

It seems appropriate to end this chapter with a quotation about the bacterial pseudovirion or generalized transducing phages:

> "It is surely one of the more bizarre manifestations of evolutionary adaptation, that a potentially lethal virus should acquire the redeeming function of a gamete, rescuing some of its victims' genes for posterity!" (Hayes, 1964).

Surely, this quotation will apply also to pseudovirions of animal cells if eventually they are shown to be capable of transferring genetic information.

7. SUMMARY

Infection of the appropriate animal cells with polyoma or SV40 virus, as well as infection of certain bacterial cells with phages P1, P22, SP10 or PBS-1, results in the production of pseudovirions. These virus-like particles contain fragments of host DNA encapsidated within viral coats. They do not contain detectable amounts of virus-specific DNA. The fragments of host DNA that are encapsidated are derived from cellular DNA synthesized before and after infection. The enzymes responsible for cleaving host DNA into pseudoviral DNA of specific size are unknown. It appears that the host DNA fragments compete with virus-specific DNA for assembly into complete particles. Virtually nothing is known, at the experimental level, about this assembly process. This is true especially for the animal viruses.

In the polyoma and SV40 systems, the host cell has an influence on the amount of pseudovirions relative to infectious virions that is produced. This host cell effect is most apparent in the SV40 system. While pseudovirions are produced after SV40 infection of primary African green monkey cells, they do not appear to be found after infection of BSC-1 cells. However, infectious SV40 virions are produced in both of these cell types.

Polyoma and SV40 pseudovirions are adsorbed to and enter a number of animal cell types. Polyoma pseudovirions are uncoated in mouse and human cells, and the pseudoviral DNA is found in the nuclear fraction. Some of the pseudoviral DNA is hydrolyzed to DNA precursors which are reutilized for DNA synthesis. As yet, attempts to show that polyoma or SV40 pseudovirions are capable of transferring specific genetic information have not been successful.

Pseudovirions of TMV have been reported. These particles contain fragments of host RNA.

Orphan pseudovirions are found in bacterial lysates. These particles contain fragments of bacterial DNA and appear to be devoid of the DNA of any known temperate or virulent phage. However, the original viral genome that directs the synthesis of the protein coat of these particles is unknown. The orphan pseudovirions of *B. subtilis* have been most extensively studied and characterized. In general, they are unable to inject their DNA into cells to which they are adsorbed. The 5′ termini of the DNA of some of these orphan pseudovirions have a high predominance of particular pyrimidines and purines. This suggests that the orphan pseudoviral DNA is cleaved from bacterial DNA by some highly specific process.

The biological role of pseudovirions is essentially unknown. However, the bacterial pseudovirions, which are more commonly known as generalized transducing particles, are capable of transferring genetic information. This transfer may have some limited evolutionary importance.

ACKNOWLEDGMENTS

I am indebted to Esther Sherberg for her patient and painstaking efforts in typing the innumerable drafts of the manuscript, checking each of the citations in the bibliography with the original publications, and other editorial assistance.

The manuscript was prepared at a time when my research was supported by grants from the following organizations: National Institute of General Medical Sciences, National Cancer Institute, and The John A. Hartford Foundation, Inc.

8. REFERENCES

Aposhian, H. V., 1970, The use of DNA for gene therapy—the need, experimental approach and implications, *Perspect. Biol. Med.* **14,** 98.

Aposhian, H. V., Qasba, P. K., Osterman, J. V., and Waddell, A., 1972, Polyoma pseudovirions: An experimental model for the development of DNA for gene therapy, *Fed. Proc.* **31,** 1310.

Aposhian, H. V., Qasba, P. K., Yelton, D. B., Pletsch, Q. A., and Sethi, V. S., 1973, The polyoma pseudovirion—its properties, production, and use in transferring DNA to mouse and human cells, *in* "Cellular Modification and Genetic Transformation by Exogenous Nucleic Acids" (R. F. Beers, Jr. and R. C. Tilghman, eds.), pp. 35–45, Johns Hopkins University Press, Baltimore.

Axelrod, D., and Trilling, D., 1972, Interaction of pseudovirions with infectious virions of Papova viruses in susceptible host cells, *Adv. Biosci.* **8,** 175–186.

Babiuk, L. A., and Hudson, J. B., 1972, 'Integration' of polyoma virus DNA into mammalian genomes, *Biochem. Biophys. Res. Commun.* **47,** 111.

Basilico, C., and Burstin, S. J., 1971, Multiplication of polyoma virus in mouse–hamster somatic hybrids: A hybrid cell line which produces viral particles containing predominantly host deoxyribonucleic acid, *J. Virol.* **7,** 802.

Basilico, C., Marin, G., and di Mayorca, G., 1966, Requirement for the integrity of the viral genome for the induction of host DNA synthesis by polyoma virus, *Proc. Natl. Acad. Sci. USA* **56,** 208.

Ben-Porat, T., and Kaplan, A. S., 1967, Correlation between replication and degradation of cellular DNA in polyoma virus-infected cells, *Virology* **32,** 457.

Blackstein, M. E., Stanners, C. P., and Farmilo, A. J., 1969, Heterogeneity of polyoma virus DNA: Isolation and characterization of non-infectious small supercoiled molecules, *J. Mol. Biol.* **42,** 301.

Bradley, E., 1965, The isolation and morphology of some new bacteriophages specific for *Bacillus* and *Acetobacter* species, *J. Gen. Microbiol.* **41,** 233.

Britten, R. J., and Kohne, D. E., 1968, Repeated sequences in DNA, *Science* (*Wash., D.C.*) **161,** 529.

Campbell, A., 1964, Transduction, *in* "The Bacteria" (I. C. Gunsalus and R. Y. Stanier, eds.), Vol. 5, pp. 49–85, Academic Press, New York.

Crawford, L. V., 1969*a*, Purification of polyoma virus, *in* "Fundamental Techniques of Virology" (K. Habel and N. P. Salzman, eds.), pp. 75–81, Academic Press, New York.

Crawford, L. V., 1969*b*, Nucleic acids of tumor viruses, *Adv. Virus Res.* **14,** 89.

Crawford, L. V., and Black, P. H., 1964, The nucleic acid of simian virus 40, *Virology* **24,** 388.

Crawford, L. V., Crawford, E. M., and Watson, D. H., 1962, The physical characteristics of polyoma virus. I. Two types of particle, *Virology* **18,** 170.

Dulbecco, R., 1969, Cell transformation by viruses: Two minute viruses are powerful tools for analyzing the mechanism of cancer, *Science* (*Wash., D.C.*) **166,** 962.

Dulbecco, R., and Vogt, M., 1963, Evidence for a ring structure of polyoma virus DNA, *Proc. Natl. Acad. Sci. USA* **50,** 236.

Dulbecco, R., Hartwell, L. H., and Vogt, M., 1965, Induction of cellular DNA synthesis by polyoma virus, *Proc. Natl. Acad. Sci. USA* **53,** 403.

Ebel-Tsipis, J., Botstein, D., and Fox, M. S., 1972*a*, Generalized transduction by phage P22 in *Salmonella typhimurium*. I. Molecular origin of transducing DNA, *J. Mol. Biol.* **71,** 433.

Ebel-Tsipis, J., Fox, M. S., and Botstein, D., 1972*b*, Generalized transduction by bacteriophage P22 in *Salmonella typhimurium*. II. Mechanism of integration of transducing DNA, *J. Mol. Biol.* **71,** 449.

Eddy, B. E., Stewart, S. E., Young, R., and Mider, G. B., 1958, Neoplasms in hamsters induced by mouse tumor agent passed in tissue culture, *J. Natl. Cancer Inst.* **20,** 747.

Eiserling, F., 1964, *Bacillus subtilis* bacteriophages: Structure, intracellular development and conditions of lysogeny, Ph.D. Thesis, University of California, 127 pp.

Fried, M., 1974, Isolation and partial characterization of different defective DNA molecules derived from polyoma virus, *J. Virol.* **13,** 939.

Gardner, S. D., Field, A. M., Coleman, D. V., and Hulme, B., 1971, New human papovavirus (B.K.) isolated from urine after renal transplantation, *Lancet* **I,** 1253.

Garro, A. J., and Marmur, J., 1970, Defective bacteriophages, *J. Cell Physiol.* **76,** 253.

Garro, A. J., Leffert, H., and Marmur, J., 1970, Genetic mapping of a defective bacteriophage on the chromosome of *Bacillus subtilis* 168, *J. Virol.* **6,** 340.

Gershon, D., Hausen, P., Sachs, L., and Winocour, E., 1965, On the mechanism of polyoma virus-induced synthesis of cellular DNA, *Proc. Natl. Acad. Sci. USA* **54,** 1584.

Gershon, D., Sachs, L., and Winocour, E., 1966, The induction of cellular DNA synthesis by simian virus 40 in contact-inhibited and in X-irradiated cells, *Proc. Natl. Acad. Sci. USA* **56**, 918.

Gilead, Z., 1972, Virus DNA and host DNA in polyoma virus-infected cells at high temperature, *Virology* **47**, 114.

Grady, L., Axelrod, D., and Trilling, D., 1970, The SV40 pseudovirus: Its potential for general transduction in animal cells, *Proc. Natl. Acad. Sci USA* **67**, 1886.

Haas, M., and Yoshikawa, H., 1969*a*, Defective bacteriophage PBSH in *Bacillus subtilis*. I. Induction, purification, and physical properties of the bacteriophage and its deoxyribonucleic acid, *J. Virol.* **3**, 233.

Haas, M., and Yoshikawa, H., 1969*b*, Defective bacteriophage PBSH in *Bacillus subtilis*. II. Intracellular development of the induced prophage, *J. Virol.* **3**, 248.

Haas, M., and Yoshikawa, H., 1969*c*, Defective bacteriophage PBSH in *Bacillus subtilis*. III. Properties of *adenine-16* marker in purified bacteriophage deoxyribonucleic acid, *J. Virol.* **4**, 844.

Hatanaka, M., and Dulbecco, R., 1966, Induction of DNA synthesis by SV40, *Proc. Natl. Acad. Sci. USA* **56**, 736.

Hayes, W., 1964, "The Genetics of Bacteria and their Viruses," John Wiley & Sons, New York.

Hirt, B., 1967, Selective extraction of polyoma DNA from infected mouse cell cultures, *J. Mol. Biol.* **26**, 365.

Hirt, B., and Widmer-Favre, C., 1972, Expression and integration of viral DNA in animal cells, *Adv. Biosci.* **8**, 167–174.

Huang, A. S., 1973, Defective interfering viruses, *Annu. Rev. Microbiol.* **27**, 101.

Huang, W. M., and Marmur, J., 1970*a*, The 5′-ends of the DNA of defective bacteriophages of *Bacillus subtilis*, *J. Mol. Biol.* **47**, 591.

Huang, W. M., and Marmur, J., 1970*b*, Characterization of inducible bacteriophages in *Bacillus licheniformis*, *J. Virol.* **5**, 237.

Ikeda, H., and Tomizawa, J., 1965*a*, Transducing fragments in generalized transduction by phage P1. I. Molecular origin of the fragments, *J. Mol. Biol.* **14**, 85.

Ikeda, H., and Tomizawa, J., 1965*b*, Transducing fragments in generalized transduction by phage P1. II. Association of DNA and protein in the fragments, *J. Mol. Biol.* **14**, 110.

Ionesco, H., Ryter, A., and Schaeffer, P., 1964, Sur un bactériophage hébergé par la souche Marburg de *Bacillus subtilis*, *Ann. Inst. Pasteur* (Paris) **107**, 764.

Kaplan, J. C., Wilbert, S. M., and Black, P. H., 1972, Endonuclease activity associated with purified simian virus 40 virions, *J. Virol.* **9**, 800.

Kasamaki, A., Ben-Porat, T., and Kaplan, A. S., 1968, Polyoma virus-induced release of inhibition of cellular DNA synthesis caused by iododeoxyuridine, *Nature* (*Lond.*) **217**, 756.

Kashmiri, S. V. S., and Aposhian, H. V., 1974, Polyoma pseudovirions. III. Study of the fate of pseudoviral DNA after infection of mouse cells with pseudovirions, *Proc. Natl. Acad. Sci. USA*, **71**, 3834.

Kidwell, W. R., Saral, R., Martin, R. G., and Ozer, H. L., 1972, Characterization of an endonuclease associated with simian virus 40 virions, *J. Virol.* **10**, 410.

Kit, S., Dubbs, D. R., and Frearson, P. M., 1966, Enzymes of nucleic acid metabolism in cells infected with polyoma virus, *Cancer Res.* **26** (Part I), 638.

Kit, S., de Torres, R. A., Dubbs, D. R., and Salvi, M. L., 1967, Induction of cellular deoxyribonucleic acid synthesis by simian virus 40, *J. Virol.* **1**, 738.

Lavi, S., and Winocour, E., 1972, Acquisition of sequences homologous to host deoxyribonucleic acid by closed circular simian virus 40 deoxyribonucleic acid, *J. Virol.* **9,** 309.

Lavi, S., and Winocour, E., 1974, Accumulation of closed-circular polyoma DNA molecules containing host DNA during serial passage of the virus, *Virology* **57,** 296.

Lavi, S., Rozenblatt, S., Singer, M. F., and Winocour, E., 1973, Acquisition of sequences homologous to host DNA by closed circular simian virus 40 DNA. II. Further studies on the serial passage of virus clone, *J. Virol.* **12,** 492.

Lennox, E. S., 1955, Transduction of linked genetic characters of the host by bacteriophage P1, *Virology* **1,** 190.

Levine, A. J., and Teresky, A. K., 1970, Deoxyribonucleic acid replication in simian virus 40-infected cells. II. Detection and characterization of simian virus 40 pseudovirions, *J. Virol.* **5,** 451.

Mattern, C. F. T., Takemoto, K. K., and DeLeva, A. M., 1967, Electron microscopic observations on multiple polyoma virus-related particles, *Virology* **32,** 378.

Michel, M. R., Hirt, B., and Weil, R., 1967, Mouse cellular DNA enclosed in polyoma viral capsids (pseudovirions), *Proc. Natl. Acad. Sci. USA* **58,** 1381.

Minowada, J., and Moore, G. E., 1963, DNA synthesis in X-irradiated cultures infected with polyoma virus, *Exp. Cell Res.* **29,** 31.

Morse, M. L., Lederberg, E. M., and Lederberg, J., 1956, Transduction in *Escherichia coli* K-12, *Genetics* **41,** 142.

Munyon, W., Kraiselburd, E., Davis, D., and Mann, J., 1971, Transfer of thymidine kinase to thymidine kinaseless L cells by infection with ultraviolet-irradiated herpes simplex virus, *J. Virol.* **7,** 813.

Okamoto, K., Mudd, J. A., Mangan, J., Huang, W. M., Subbaiah, T. V., and Marmur, J., 1968*a*, Properties of the defective phage of *Bacillus subtilis*, *J. Mol. Biol.* **34,** 413.

Okamoto, K., Mudd, J. A., and Marmur, J., 1968*b*, Conversion of *Bacillus subtilis* DNA to phage DNA following mitomycin C induction, *J. Mol. Biol.* **34,** 429.

Okubo, S., Stodolsky, M., Bott, K., and Strauss, B., 1963, Separation of the transforming and viral deoxyribonucleic acids of a transducing bacteriophage of *Bacillus subtilis*, *Proc. Natl. Acad. Sci. USA* **50,** 679.

Osterman, J. V., Waddell, A., and Aposhian, H. V., 1970, DNA and gene therapy: Uncoating of polyoma pseudovirus in mouse embryo cells, *Proc. Natl. Acad. Sci. USA* **67,** 37.

Ottolenghi-Nightingale, E., 1969, Induction of melanin synthesis in albino mouse skin by DNA from pigmented mice, *Proc. Natl. Acad. Sci. USA* **64,** 184.

Ozeki, H., and Ikeda, H., 1968, Transduction mechanisms, *in Annu. Rev. Genet.* **2,** 245–278.

Padgett, B. L., Walker, D. L., ZuRhein, G. M., and Eckroade, R. J , 1971, Cultivation of papova-like virus from human brain with progressive multifocal leucoencephalopathy, *Lancet* **I,** 1257.

Qasba, P. K., and Aposhian, H. V., 1971, DNA and gene therapy: Transfer of mouse DNA to human and mouse embryonic cells by polyoma pseudovirions, *Proc. Natl. Acad. Sci. USA* **68,** 2345.

Qasba, P. K., Yelton, D. B., Pletsch, Q. A., and Aposhian, H. V., 1974, *in* "Molecular Studies in Viral Neoplasia, M. D. Anderson Annual Symposium on Fundamental Cancer Research, 1972," pp. 169–189, Williams & Wilkins Co., Baltimore.

Ralph, R. K., and Colter, J. S., 1972, Evidence for the integration of polyoma virus DNA in a lytic system, *Virology* **48,** 49.

Ritzi, E., and Levine, A. J., 1970, Deoxyribonucleic acid replication in simian virus 40-

infected cells. III. Comparison of simian virus 40 lytic infection in three different monkey kidney cell lines, *J. Virol.* **5,** 686.
Ritzi, E. M., and Levine, A. J., 1973, The fragmentation of cellular DNA and the formation of pseudovirions during SV 40 infection of African green monkey kidney cells, *J. Gen. Virol.* **20,** 353.
Rogers, S., and Moore, M., 1963, Studies of the mechanism of action of the Shope rabbit papilloma virus. I. Concerning the nature of the induction of arginase in the infected cells, *J. Exp. Med.* **117,** 521.
Rozenblatt, S., Lavi, S., Singer, M. F., and Winocour, E., 1973, Acquisition of sequences homologous to host DNA by closed circular simian virus 40 DNA. III. Host sequences, *J. Virol.* **12,** 501.
Seaman, E., Tarmy, E., and Marmur, J., 1964, Inducible phages of *Bacillus subtilis*, *Biochemistry* **3,** 607.
Sheppard, J. R., Levine, A. J., and Burger, M. M., 1971, Cell-surface changes after infection with oncogenic viruses: requirement for synthesis of host DNA, *Science* (*Wash., D.C.*) **172,** 1345.
Siegel, A., 1971, Pseudovirions of tobacco mosaic virus, *Virology* **46,** 50.
Siegel, E. C., and Marmur, J., 1969, Temperature-sensitive induction of bacteriophages in *Bacillus subtilis* 168, *J. Virol.* **4,** 610.
Skoog, L., Nordenskjöld, B. A., and Lindberg, U., 1970, Deoxyribonucleotide pools and deoxyribonucleic acid synthesis in mouse embryo cells infected with three classes of polyoma virus particles, *J. Virol.* **6,** 28.
Sokol, F., Huebner, F. K., Tan, K. B., Sheehan, M. C., and Santoli, D., 1974, Heterogeneity of simian virus 40 population: Virus particles containing one or two virus DNA molecules? *J. Gen. Virol.* **22,** 55.
Stewart, S. E., Eddy, B. E., Gochenour, A. M., Borgese, N. G., and Grubbs, G. E., 1957, The induction of neoplasms with a substance released from mouse tumors by tissue culture, *Virology* **3,** 380.
Stickler, D. J., Tucker, R. G., and Kay, D., 1965, Bacteriophage-like particles released from *Bacillus subtilis* after induction with hydrogen peroxide, *Virology* **26,** 142.
Stocker, B. A. D., 1958, Phage-mediated transduction, *in* "Recent Progress in Microbiology" (G. Tunevall, ed.), Vol. VII, International Congress for Microbiology, pp. 31–39, Charles C. Thomas, Springfield, Ill.
Streisinger, G., Emrich, J., and Stahl, M. M., 1967, Chromosome structure in phage T4. III. Terminal redundancy and length determination, *Proc. Natl. Acad. Sci. USA* **57,** 292.
Tai, H. T., Smith, C. A., Sharp, P. A., and Vinograd, J., 1972, Sequence heterogeneity in closed simian virus 40 deoxyribonucleic acid, *J. Virol.* **9,** 317.
Tegtmeyer, P., 1972, Simian virus 40 deoxyribonucleic acid synthesis: The viral replicon, *J. Virol.* **10,** 591.
Tegtmeyer, P., and Ozer, H. L., 1971, Temperature-sensitive mutants of simian virus 40: Infection of permissive cells, *J. Virol.* **8,** 516.
Todaro, G. J., Lazar, G. K., and Green, H., 1965, The initiation of cell division in a contact-inhibited mammalian cell line, *J. Cell. Comp. Physiol.* **66,** 325.
Tooze, J., 1973, "The Molecular Biology of Tumour Viruses," Cold Spring Harbor Laboratory, Cold Spring Harbor, N. Y.
Trilling, D. M., and Axelrod, D., 1970, Encapsidation of free host DNA by simian virus 40: A simian virus 40 pseudovirus, *Science* (*Wash., D.C.*) **168,** 268.
Türler, H., 1974*a*, Interactions of polyoma and mouse DNAs. I. Lytic infection of bromodeoxyuridine-prelabeled mouse embryo cells, *J. Virol.* **13,** 276.

Türler, H., 1974*b*, Interactions of polyoma and mouse DNAs. II. Polyoma-induced mouse DNA replication and pseudovirion formation, *J. Virol.* **13,** 285.

Uchida, S., Watanabe, S., and Kato, M., 1966, Incomplete growth of simian virus 40 in African green monkey kidney culture induced by serial undiluted passages, *Virology* **28,** 135.

Vesco, C., and Basilico, C., 1971, Induction of mitochondrial DNA synthesis by polyoma virus, *Nature* (*Lond.*) **229,** 336.

Vinograd, J., Lebowitz, J., Radloff, R., Watson, R., and Laipis, P., 1965, The twisted circular form of polyoma viral DNA, *Proc. Natl. Acad. Sci. USA* **53,** 1104.

Weil, R., and Vinograd, J., 1963, The cyclic helix and cyclic coil forms of polyoma viral DNA, *Proc. Natl. Acad. Sci. USA* **50,** 730.

Weil, R., Michel, M. R., and Ruschmann, G. K., 1965, Induction of cellular DNA synthesis by polyoma virus, *Proc. Natl. Acad. Sci. USA* **53,** 1468.

Weiner, L. P., Herndon, R. M., Narayan, O., Johnson, R. T., Shah, K., Rubinstein, L. J., Preziosi, T. J., and Conley, F. K., 1972, Isolation of virus related to SV40 from patients with progressive multifocal leukoencephalopathy, *New Engl. J. Med.* **286,** 385.

Westphal, H., and Dulbecco, R., 1968, Viral DNA in polyoma- and SV40-transformed cell lines, *Proc. Natl. Acad. Sci. USA* **59,** 1158.

Winocour, E., 1963, Purification of polyoma virus, *Virology* **19,** 158.

Winocour, E., 1967*a*, On the apparent homology between DNA from polyoma virus and normal mouse synthetic RNA, *Virology* **31,** 15.

Winocour, E., 1967*b*, Studies on the basis for the observed homology between DNA from polyoma virus and DNA from normal mouse cells, *in* "The Molecular Biology of Viruses" (J. S. Colter and W. Paranchych, ed.), pp. 577–591, Academic Press, New York.

Winocour, E., 1968, Further studies on the incorporation of cell DNA into polyoma-related particles, *Virology* **34,** 571.

Winocour, E., 1969, Some aspects of the interaction between polyoma virus and cell DNA, *Adv. Virus Res.* **14,** 153.

Winocour, E., Kaye, A. M., and Stollar, V., 1965, Synthesis and transmethylation of DNA in polyoma-infected cultures, *Virology* **27,** 156.

Yamagishi, H., and Takahashi, I., 1968, Transducing particles of PBS1, *Virology* **36,** 639.

Yelton, D. B., and Aposhian, H. V., 1972, Polyoma pseudovirions. I. Sequence of events in primary mouse embryo cells leading to pseudovirus production, *J. Virol.* **10,** 340.

Yelton, D. B., and Aposhian, H. V., 1973, Polyoma pseudovirions. II. Influence of host cell on pseudovirus production, *J. Virol.* **12,** 1065.

Yoshiike, K., 1968, Studies on DNA from low-density particles of SV40. I. Heterogeneous defective virions produced by successive undiluted passages, *Virology* **34,** 391.

Zinder, N. D., 1953, Infective heredity in bacteria, *Cold Spring Harbor Symp. Quant. Biol.* **18,** 261.

Zinder, N. D., 1955, Bacterial transduction, *J. Cell. Comp. Physiol.* 23. (suppl. 2), **45.**

Zinder, N. D., and Lederberg, J., 1952, Genetic exchange in *Salmonella*, *J. Bacteriol.* **64,** 679.

zur Hausen, H., 1969, Failure to detect incorporation of host cell DNA into virions of adenovirus type 12, *Virology* **38,** 194.

Index

GPSR Compliance

The European Union's (EU) General Product Safety Regulation (GPSR) is a set of rules that requires consumer products to be safe and our obligations to ensure this.

If you have any concerns about our products, you can contact us on ProductSafety@springernature.com

In case Publisher is established outside the EU, the EU authorized representative is:

Springer Nature Customer Service Center GmbH
Europaplatz 3
69115 Heidelberg, Germany

Batch number: 10615115

Printed by Printforce, the Netherlands